Numerical Solutions for Nanocomposite Structures

Numerical Solutions for Nanocomposite Structures provides an in-depth exploration of structural analysis using numerical methods grounded in rigorous mathematical modeling. Theoretical foundations are established by comprehensively elucidating theories governing beams, plates, and shells, leading to the derivation of governing equations based on the stress–strain relationship. The process of obtaining governing equations through the energy method, the application of boundary conditions, and the utilization of numerical methods to calculate deflection, frequency, and buckling loads is meticulously explained, providing readers with valuable insights into structural analysis methodologies.

- Includes diverse numerical examples involving beams, plates, and pipes, providing a comprehensive understanding of underlying theories and relationships.
- Provides numerous practical examples demonstrating the application of numerical methods to address challenges in civil and mechanical engineering problems.
- Discusses the unique mechanical, thermal, and electrical properties of nanocomposites and how they can be utilized in various industries.

Numerical Solutions for Nanocomposite Structures

Maryam Shokravi and Amin Shagholani Loor

CRC Press
Taylor & Francis Group
Boca Raton London New York

CRC Press is an imprint of the
Taylor & Francis Group, an **informa** business

Designed cover image: Shutterstock

First edition published 2025
by CRC Press
2385 NW Executive Center Drive, Suite 320, Boca Raton FL 33431

and by CRC Press
4 Park Square, Milton Park, Abingdon, Oxon, OX14 4RN

CRC Press is an imprint of Taylor & Francis Group, LLC

© 2025 Maryam Shokravi and Amin Shagholani Loor

ISBN: 978-1-032-83990-5 (hbk)
ISBN: 978-1-032-83989-9 (pbk)
ISBN: 978-1-003-51071-0 (ebk)

DOI: 10.1201/9781003510710

Typeset in Times LT Std
by Apex CoVantage, LLC

Contents

Figures

About the Author

Maryam Shokravi received her PhD degree from the University of Kashan, Iran, in the field of engineering in 2015. During her sabbatical period at Northeastern University in Boston, USA, in 2014–2015, she focused on nano and piezo-electric materials in mechanical engineering. With a master's degree in mechanical engineering from Kashan University and a bachelor's degree from Iran University of Science & Technology, Maryam has a strong educational background. She has extensive teaching experience, having served as a faculty member at Buein Zahra Technical University and Islamic Azad University of Saveh and Principal of Mehrab High School. Her courses taught include Strength of Material, Statics, Fluid Mechanics, Heat Transfer, Thermodynamics, Dynamics, Vibration, Physics, and Mathematics.

Maryam Shokravi has received several honors and awards, including the Best Researcher University Award at Buein Zahra Technical University in 2017, and ranked first in the Iranian Physics Olympiad Exam among about 35,000 students in 1996. Her research areas encompass applied mechanics, computational solid mechanics, mathematical modeling, numerical solutions of partial differential equations, modeling of micro/nano structures, higher-order continuum theories, piezo-electric materials, and functionally graded materials.

Maryam has an impressive list of publications in reputable journals and conferences, covering topics such as vibrations, buckling, and dynamic analyses of various structures, including nanocomposites, sandwich plates, and cylindrical shells. She has contributed to advancing the understanding of complex phenomena in the field of mechanics and materials.

Amin Shagholani Loor is a dedicated civil engineer with a wealth of experience and expertise. He received his PhD degree from Khomein Azad University in 2023. Amin has already earned a master's degree from East Azerbaijan University of Science and Research and a bachelor's degree from Anzali Azad University. Throughout his impressive career, Amin has served as the Mayor of Dillman, a university lecturer for 12 years in associate and bachelor's degree programs in civil engineering, and the Construction Manager of Siahkol municipality for five years. Amin's multifaceted role includes responsibilities such as controlling structural drawings, supervising construction projects, and actively participating in various engineering associations.

Amin has showcased his expertise through published papers in reputable journals, focusing on the reliability analysis and optimization of building beams reinforced by steel fibers. Additionally, he has actively participated in seminars and workshops, contributing to his deep understanding of engineering rights and the interpretation of regulations.

With a proven track record of designing, supervising, and implementing a wide range of residential and commercial buildings, Amin Shagholani Loor stands out as a versatile and accomplished professional in the field of civil engineering dedicated to advancing the industry through his expertise and commitment to ongoing learning and innovation.

Foreword

In this book, an in-depth exploration of structural analysis is undertaken through the lens of numerical methods grounded in rigorous mathematical modeling. The analysis encompasses a detailed examination of beam, plate, and pipe elements, providing insights into their deflection, frequency responses, and buckling loads. The theoretical foundation is laid by comprehensively elucidating various theories governing beams, plates, and shells. The intricate relationship between stress and strain is then articulated, leading to the derivation of governing equations for the structures under consideration. To address these governing equations using numerical methods, two broad categories are discussed: those based on derivatives and those based on integrals. Derivative-based methods, notably the differential quadrature method, are highlighted for their flexibility in modeling boundary conditions, efficient analysis time, and remarkable accuracy.

The book culminates in the presentation of diverse numerical examples involving beams, plates, and pipes, offering a holistic understanding of the underlying theories and relationships. The process of obtaining governing equations through the energy method, the application of boundary conditions, and the utilization of numerical methods to calculate deflection, frequency, and buckling loads is meticulously explained, providing readers with valuable insights into structural analysis methodologies.

Preface

The mathematical modeling of nanocomposite structures and the pursuit of numerical solutions represent a pivotal frontier in materials science and engineering. Nanocomposites, characterized by the incorporation of nanoscale reinforcements into a matrix material, exhibit unique mechanical, thermal, and electrical properties, offering a broad spectrum of applications in various industries. The intricate interplay of nanoscale phenomena demands advanced mathematical models to accurately capture the behavior of these structures under different conditions. The development of robust numerical solutions is essential for predicting and understanding the complex responses of nanocomposite materials. This field not only explores the fundamental principles governing nanocomposite behavior but also addresses practical engineering challenges. As researchers delve into this subject, they contribute to the advancement of innovative materials with tailored properties, influencing the design and optimization of nanocomposite-based technologies across diverse domains, including aerospace, electronics, and healthcare. The importance of this research lies in its potential to unlock new frontiers in material science, enabling the creation of advanced materials with enhanced performance characteristics and expanded applications.

Acknowledgments

The completion of this book has been made possible through the collective effort and support of numerous individuals, whose contributions, whether direct or indirect, have been invaluable.

We extend my heartfelt appreciation to all those who have provided guidance, encouragement, and assistance throughout the journey of writing and publishing this book.

To the editors, advisors, and reviewers whose expertise and feedback have helped shape the content and structure of this work, I am deeply grateful.

I am indebted to the countless researchers, scholars, and creators whose work has laid the foundation for the ideas presented in this book. It is very important for us to thank Artin Ketabdar and Amir Mohammad Zoghi for helping to write this book. Your contributions to the field are recognized and appreciated.

My gratitude also extends to my family, friends, and colleagues for their unwavering support, understanding, and encouragement during the process of bringing this book to fruition.

Finally, to the readers who will engage with this book, I thank you for your interest and trust. It is my sincere hope that this work will resonate with you and contribute meaningfully to your understanding or enjoyment.

Thank you to all who have played a part, whether big or small, in making this book a reality.

1 Mathematical Modeling of Structures

1.1 INTRODUCTION

Structural engineering forms the backbone of modern infrastructure, encompassing a diverse array of architectural marvels, from towering skyscrapers to intricate bridges and resilient industrial facilities. At the heart of understanding and designing these structures lies the intricate realm of mathematical modeling. "Mathematical Modeling of Structures" serves as a comprehensive guide, navigating the intricate landscape of structural analysis through mathematical frameworks and theoretical constructs.

In this introduction, we embark on a journey into the realm of structural modeling, exploring the fundamental principles that underpin the behavior of beams, plates, and shells. By leveraging mathematical tools and theoretical frameworks, engineers and researchers gain insight into the complex interplay of forces and deformations that govern structural behavior.

Throughout this text, we delve into the essence of strain relations and assumptions, laying the groundwork for understanding nonlinear strain behaviors and their implications in structural analysis. The mathematical modeling of beams takes center stage, with an in-depth exploration of classical theories such as Euler–Bernoulli and Timoshenko, alongside more advanced formulations including sinusoidal, hyperbolic, and exponential shear deformation theories.

Transitioning from beams to plates and shells, the text traverses classical and modern modeling approaches, illuminating the intricacies of first-order theories, Reddy's formulations, and sinusoidal shear deformation theories. By synthesizing these diverse theories, researchers and practitioners gain a holistic understanding of structural behavior across a spectrum of geometries and loading conditions.

As we embark on this exploration of mathematical modeling in structural engineering, it is imperative to recognize the profound impact that these methodologies have on shaping our built environment. By honing our understanding of structural behavior through mathematical rigor, we empower ourselves to design safer, more efficient, and more resilient structures that stand the test of time. Join us as we delve into the rich tapestry of mathematical modeling in structural engineering, unraveling the complexities of structural behavior and paving the way for innovation and excellence in design and analysis.

1.2 STRAIN RELATIONS

In the realm of structural engineering, understanding the relationship between strains and deformations is paramount for accurate analysis and design. Strain, defined as

DOI: 10.1201/9781003510710-1

the measure of deformation experienced by a material subjected to external forces, serves as a fundamental parameter in quantifying the responses of structures to loads. Strain relations form the cornerstone of structural modeling, providing insights into how materials respond to stress and how deformations manifest within structural elements. These relations are often derived from the underlying assumptions of the chosen theoretical framework, guiding the mathematical representation of structural behavior.

Central to many strain relations is the concept of linear elasticity, where deformations are assumed to be proportional to applied loads within the elastic limit of a material. However, as structures experience higher loads or exhibit nonlinear behaviors, more sophisticated strain relations become necessary to accurately capture their response. Nonlinear strain relations, such as those based on von Karman assumptions, account for the effects of large deformations and geometric nonlinearity. These formulations consider additional factors such as strain–displacement relationships and material nonlinearity, offering a more comprehensive depiction of structural behavior under varying conditions.

Furthermore, the choice of strain relation is often influenced by the specific geometry and loading conditions of the structure under consideration. Different theories may be employed for beams, plates, and shells, each tailored to capture the unique modes of deformation exhibited by these structural elements. In essence, strain relations serve as the bridge between applied loads and resulting deformations, providing engineers with the necessary tools to predict and analyze the behavior of structures with accuracy and precision. By understanding the intricate interplay between strains and deformations, engineers can optimize designs, enhance structural performance, and ensure the safety and reliability of built environments. The subsequent equation can be expressed as follows [1, 2]:

$$\varepsilon_{ij} = \frac{1}{2}\left(\sum_{i=1}^{3}\frac{\partial y_i}{\partial Y_j}\frac{\partial y_i}{\partial Y_j} - \delta_{ij}\right),\tag{1.1}$$

where:

$$\varepsilon_{11} = \frac{\partial u_1}{\partial x_1} + \frac{1}{2}\left[\left(\frac{\partial u_1}{\partial x_1}\right)^2 + \left(\frac{\partial u_2}{\partial x_1}\right)^2 + \left(\frac{\partial u_3}{\partial x_1}\right)^2\right],\tag{1.2}$$

$$\varepsilon_{22} = \frac{\partial u_2}{\partial x_2} + \frac{1}{2}\left[\left(\frac{\partial u_1}{\partial x_2}\right)^2 + \left(\frac{\partial u_2}{\partial x_2}\right)^2 + \left(\frac{\partial u_3}{\partial x_2}\right)^2\right],\tag{1.3}$$

$$\varepsilon_{33} = \frac{\partial u_3}{\partial x_3} + \frac{1}{2}\left[\left(\frac{\partial u_1}{\partial x_3}\right)^2 + \left(\frac{\partial u_2}{\partial x_3}\right)^2 + \left(\frac{\partial u_3}{\partial x_3}\right)^2\right],\tag{1.4}$$

$$\varepsilon_{12} = \frac{1}{2}\left(\frac{\partial u_1}{\partial x_2} + \frac{\partial u_2}{\partial x_1} + \frac{\partial u_1}{\partial x_1}\frac{\partial u_1}{\partial x_2} + \frac{\partial u_2}{\partial x_1}\frac{\partial u_2}{\partial x_2} + \frac{\partial u_3}{\partial x_1}\frac{\partial u_3}{\partial x_2}\right), \tag{1.5}$$

$$\varepsilon_{13} = \frac{1}{2}\left(\frac{\partial u_1}{\partial x_3} + \frac{\partial u_3}{\partial x_1} + \frac{\partial u_1}{\partial x_1}\frac{\partial u_1}{\partial x_3} + \frac{\partial u_2}{\partial x_1}\frac{\partial u_2}{\partial x_3} + \frac{\partial u_3}{\partial x_1}\frac{\partial u_3}{\partial x_3}\right), \tag{1.6}$$

$$\varepsilon_{23} = \frac{1}{2}\left(\frac{\partial u_2}{\partial x_3} + \frac{\partial u_3}{\partial x_2} + \frac{\partial u_1}{\partial x_2}\frac{\partial u_1}{\partial x_3} + \frac{\partial u_2}{\partial x_2}\frac{\partial u_2}{\partial x_3} + \frac{\partial u_3}{\partial x_2}\frac{\partial u_3}{\partial x_3}\right). \tag{1.7}$$

The Green–Lagrange strain tensor is a fundamental concept in continuum mechanics used to quantify deformations in materials subjected to external forces. Named after the mathematicians George Green and Joseph Louis Lagrange, this strain tensor provides a measure of strain that accounts for both linear and nonlinear deformations, making it particularly useful for analyzing materials under large deformations.

In its essence, the Green–Lagrange strain tensor is a symmetric second-order tensor that describes the change in shape and size of material particles within a deformable body. The first two terms in the expression represent the linear strain components, while the third term accounts for the nonlinear or geometric strain components. This third term arises due to the fact that under large deformations, the strains depend not only on the displacements themselves but also on their gradients.

One of the key advantages of the Green–Lagrange strain tensor is its ability to capture the effects of large deformations and nonlinear material behavior. Unlike other strain measures such as engineering strain or Cauchy strain, which are only accurate for small deformations, the Green–Lagrange strain tensor remains valid even under significant distortions. The Green–Lagrange strain tensor is widely used in various fields of engineering and physics, including solid mechanics, materials science, and computational modeling. It serves as a foundational concept for understanding and quantifying the behavior of materials subjected to complex loading conditions, providing engineers and researchers with a powerful tool for predicting structural responses and designing innovative solutions.

1.3 BEAM THEORIES

1.3.1 EULER–BERNOULLI THEORY

The Euler–Bernoulli beam theory is a fundamental framework used to model the behavior of beams under bending loads. Developed by Leonhard Euler and Daniel Bernoulli in the 18th century, this theory provides a simplified yet powerful approach to analyzing the deformation and stresses within beams. At its core, the Euler–Bernoulli theory assumes that beams undergo small deformations and that cross-sections perpendicular to the neutral axis remain plane and perpendicular to the longitudinal axis of the beam. This assumption allows for the derivation of a linear differential equation governing the relationship between bending moment, curvature, and applied loads along the length of the beam.

The Euler–Bernoulli theory provides a simple yet accurate approximation for beams with relatively slender geometries and low aspect ratios, making it a widely used tool in engineering design and analysis. Despite its simplicity, the Euler–Bernoulli theory has limitations, particularly when applied to beams with significant shear deformations or non-uniform cross-sections. In such cases, more advanced theories like the Timoshenko beam theory or finite element analysis may be necessary to provide a more accurate representation of the beam's behavior. However, for many engineering applications, the Euler–Bernoulli theory remains a valuable and practical tool for understanding and predicting the response of beams to bending loads. Based on this theory, we have [2]:

$$u_x(x,z,t) = u(x,t) - z\frac{\partial w(x,t)}{\partial x}, \tag{1.8}$$

$$u_y(x,z,t) = 0, \tag{1.9}$$

$$u_z(x,z,t) = w(x,t). \tag{1.10}$$

The nonlinear strain-displacement relation for this theory is:

$$\varepsilon_{xx} = \left(\frac{\partial u}{\partial x}\right) + \frac{1}{2}\left(\frac{\partial w}{\partial x}\right)^2 - z\left(\frac{\partial^2 w}{\partial x^2}\right). \tag{1.11}$$

1.3.2 TIMOSHENKO BEAM THEORY

The Timoshenko beam theory, developed by Stephen Timoshenko in the early 20th century, extends the classical Euler–Bernoulli beam theory by accounting for shear deformations and rotational effects. Unlike the Euler–Bernoulli theory, which assumes that shear deformations are negligible and that cross-sections remain perpendicular to the neutral axis during bending, the Timoshenko theory acknowledges the significant influence of shear deformation in beams, especially those with short spans or high aspect ratios. In the Timoshenko beam theory, the assumption of constant cross-sections perpendicular to the neutral axis is relaxed, allowing for warping and twisting of the cross-sections during bending. This consideration is particularly important for beams with open or thin-walled cross-sections, where shear deformation effects are more pronounced.

By incorporating shear deformation effects, the Timoshenko beam theory provides a more accurate representation of the behavior of beams subjected to bending loads. It predicts higher deflections and more realistic stress distributions compared to the Euler–Bernoulli theory, especially near the supports and at points of concentrated loads. Consequently, the Timoshenko theory is often preferred for analyzing beams with short spans, thick cross-sections, or non-uniform loading conditions. In engineering practice, the Timoshenko beam theory serves as a valuable tool for designing and analyzing a wide range of structures, including beams, frames, and bridges. By accounting for shear deformations and rotational effects, it offers engineers a more comprehensive understanding of beam behavior, enabling them to design safer and more efficient structures that meet the demands of modern construction. Based on this theory, we have [3]:

$$u_x(x,z,t) = u(x,t) + z\psi(x,t), \tag{1.12}$$

$$u_y(x,z,t) = 0, \tag{1.13}$$

$$u_z(x,z,t) = w(x,t), \tag{1.14}$$

where ψ is the rotation. The nonlinear strain–displacement relations for this theory are:

$$\varepsilon_{xx} = \frac{\partial u}{\partial x} + z\frac{\partial \psi}{\partial x} + \frac{1}{2}\left(\frac{\partial w}{\partial x}\right)^2, \tag{1.15}$$

$$\varepsilon_{xz} = \frac{\partial w}{\partial x} + \psi. \tag{1.16}$$

1.3.3 SINUSOIDAL SHEAR DEFORMATION THEORY

The sinusoidal shear deformation beam theory is a refined approach for modeling the behavior of beams subjected to bending loads. Unlike classical beam theories such as the Euler–Bernoulli and Timoshenko theories, which assume constant shear deformation distributions along the beam's cross-section, the sinusoidal shear deformation theory considers a sinusoidal variation in shear strains across the thickness of the beam.

By incorporating this sinusoidal variation, the theory captures more accurately the non-uniform distribution of shear stresses and strains within the beam, especially in cases where the beam's cross-section is not symmetrical or varies along its length. This allows for a more realistic representation of shear deformations, particularly in beams with thick or non-uniform cross-sections. One of the key advantages of the sinusoidal shear deformation theory is its ability to provide accurate predictions of beam behavior, even for complex loading and boundary conditions. It offers engineers a refined tool for analyzing the responses of beams in structural systems, enabling more precise calculations of deflections, stresses, and critical failure modes.

In practical engineering applications, the sinusoidal shear deformation theory finds utility in the design and analysis of various structures, including bridges, buildings, and aerospace components. By accounting for the sinusoidal variation in shear deformations, engineers can ensure the integrity and reliability of their designs, lead-ing to safer and more efficient structures in diverse engineering applications. Based on this theory, we have [4]:

$$u_x(x,z,t) = u(x,t) - z\frac{\partial w}{\partial x} + \underbrace{\left(h/\pi \sin\left(\pi z/h\right)\right)}_{f(z)}\psi(x,t), \tag{1.17}$$

$$u_y(x,z,t) = 0, \tag{1.18}$$

$$u_z(x,z,t) = w(x,t), \tag{1.19}$$

where ψ is the rotation. The nonlinear strain–displacement relations for this theory are:

$$\varepsilon_{xx} = \frac{\partial u}{\partial x} - z\frac{\partial^2 w}{\partial x^2} + \frac{1}{2}\left(\frac{\partial w}{\partial x}\right)^2 + f\frac{\partial \psi}{\partial x}, \tag{1.20}$$

$$\varepsilon_{xz} = \underbrace{\cos\left(\frac{\pi z}{h}\right)}_{\partial f/\partial z}\psi. \tag{1.21}$$

1.3.4 Hyperbolic Shear Deformation Theory

The hyperbolic shear deformation theory is an advanced approach used in struc-tural analysis to model the behavior of beams under bending loads. Unlike tradi-tional beam theories such as the Euler–Bernoulli and Timoshenko theories, which assume constant shear deformation distributions across the beam's cross-section, the hyperbolic shear deformation theory considers a hyperbolic variation in shear strains through the beam's thickness. This theory accounts for the non-uniform distribution of shear stresses and strains within the beam, particularly in cases where the beam's cross-section is asymmetrical or varies along its length. By incorporating this hyper-bolic variation, the theory provides a more accurate representation of shear deforma-tions, especially in beams with thick or non-uniform cross-sections.

One of the main advantages of the hyperbolic shear deformation theory is its ability to capture more complex deformation patterns, leading to improved accuracy in predicting beam behavior. Engineers can use this theory to obtain more precise calculations of deflections, stresses, and failure modes, particularly in structures subjected to significant bending loads or with challenging geometries. In engineer-ing practice, the hyperbolic shear deformation theory is applied in various fields, including aerospace engineering, civil engineering, and mechanical engineering. Its advanced capabilities make it a valuable tool for designing and analyzing a wide range of structures, from aircraft wings and rotor blades to bridges and industrial machinery. By considering the hyperbolic variation in shear deformations, engineers can develop safer, more efficient designs that meet the demands of modern engineering challenges. Based on this theory, we have [4]:

$$u_x(x,z,t) = u(x,t) - z\frac{\partial w(x,t)}{\partial x} + \underbrace{\left[h\sin h\left(\frac{z}{h}\right) - z\cos h\left(\frac{1}{2}\right)\right]}_{\Phi(z)}\left(\frac{\partial w(x,t)}{\partial x} - \psi(x,t)\right), \quad (1.22)$$

$$u_y(x,z,t) = 0, \quad\quad\quad (1.23)$$

$$u_z(x,z,t) = w(x,t), \quad\quad\quad (1.24)$$

The nonlinear strain-displacement relations for this theory are:

$$\varepsilon_{xx} = \left(\frac{\partial u}{\partial x}\right) - z\left(\frac{\partial^2 w}{\partial x^2}\right) + \frac{1}{2}\left(\frac{\partial w}{\partial x}\right)^2 + \left[h\sin h\left(\frac{z}{h}\right) - z\cos h\left(\frac{1}{2}\right)\right]\left(\frac{\partial^2 w}{\partial x^2} - \frac{\partial \psi}{\partial x}\right), \quad (1.25)$$

$$\varepsilon_{xz} = \left(\cos h(z) - \cos h\left(\frac{1}{2}\right)\right)\left(\frac{\partial w}{\partial x} - \psi\right). \quad\quad (1.26)$$

1.4 PLATE THEORIES

1.4.1 CLASSICAL PLATE THEORY

Classical plate theory, also known as Kirchhoff–Love theory, provides a fundamental framework for analyzing the behavior of thin plates subjected to loads. Developed based on the assumptions of small deflections and linear material behavior, this theory offers a simplified yet powerful approach for modeling the deformation and stresses within plates. In classical plate theory, plates are considered thin structures with negligible bending resistance in the thickness direction. As a result, the theory assumes that the plate experiences only in-plane deformations, neglecting any out-of-plane bending or stretching effects. This simplification allows for the derivation of a set of governing equations that describe the behavior of the plate under various loading conditions.

One of the key assumptions of classical plate theory is that the plate's cross-section remains planar and perpendicular to the mid-surface during deformation. This assumption is valid for thin plates with small deflections, where the effects of bending and stretching in the thickness direction are negligible compared to in-plane deformations. Classical plate theory provides engineers with a valuable tool for analyzing the behavior of thin plate structures, such as aircraft wings, ship hulls, and building facades. By considering only in-plane deformations, engineers can simplify the analysis process and obtain accurate predictions of deflections, stresses, and buckling modes within the plate.

Despite its simplicity, classical plate theory has limitations, particularly for plates subjected to large deflections or nonlinear material behavior. In such cases, more advanced theories, such as the first-order shear deformation theory or higher-order plate theories, may be necessary to provide a more accurate representation of the plate's behavior. However, for many engineering applications, classical plate theory remains a valuable and practical tool for analyzing thin plate structures and designing innovative solutions. Based on this theory, we have [1]:

$$u_x\left(x,y,z,t\right) = u\left(x,y,t\right) - z\frac{\partial w\left(x,y,t\right)}{\partial x}, \tag{1.27}$$

$$u_y\left(x,y,z,t\right) = v\left(x,y,t\right) - z\frac{\partial w\left(x,y,t\right)}{\partial y}, \tag{1.28}$$

$$u_z\left(x,y,z,t\right) = w\left(x,y,t\right). \tag{1.29}$$

The strain–displacement relations for this theory are:

$$\varepsilon_{xx} = \frac{\partial u}{\partial x} - z\frac{\partial^2 w}{\partial x^2}, \tag{1.30}$$

$$\varepsilon_{yy} = \frac{\partial v}{\partial y} - z\frac{\partial^2 w}{\partial y^2}, \tag{1.31}$$

$$\varepsilon_{xy} = \frac{1}{2}\left(\frac{\partial u}{\partial y} + \frac{\partial v}{\partial x}\right) - z\frac{\partial^2 w}{\partial x\partial y}. \tag{1.32}$$

1.4.2 First-Order Shear Deformation Plate Theory

First-order shear deformation plate theory is an advancement of classical plate theory, aiming to address some of its limitations. Unlike classical plate theory, which assumes that the cross-sections of plates remain plane and perpendicular to the mid-surface during deformation, first-order shear deformation theory considers the effects of shear deformation within the plate thickness. In this theory, the displacement field is allowed to vary linearly through the thickness of the plate, capturing the shearing effects that occur as the plate deforms. By incorporating this shear deformation, the theory provides a more accurate representation of the plate's behavior, especially for thicker plates or those subjected to significant bending loads.

One of the key advantages of the first-order shear deformation theory is its ability to predict more realistic deflections and stress distributions within the plate. By considering the variation in displacement through the thickness, engineers can obtain more accurate results for plates with non-uniform cross-sections or complex loading conditions.

In engineering practice, the first-order shear deformation theory finds widespread use in the analysis and design of thin and moderately thick plate structures.

It serves as a valuable tool for predicting deflections, stresses, and buckling modes, particularly in aerospace, civil, and mechanical engineering applications. Despite its advancements, the first-order shear deformation theory still has limitations, particularly for plates with very thick cross-sections or highly nonlinear material behavior. In such cases, higher-order plate theories or finite element analysis may be necessary to provide more accurate predictions. Nonetheless, for many engineering applications, the first-order shear deformation theory remains a practical and effective tool for analyzing plate structures and informing design decisions. Based on this theory, we have [2]:

$$u_x(x,y,z,t) = u(x,y,t) + z\phi_x(x,y,t), \tag{1.33}$$

$$u_y(x,y,z,t) = v(x,y,t) + z\phi_y(x,y,t), \tag{1.34}$$

$$u_z(x,y,z,t) = w(x,y,t), \tag{1.35}$$

where ϕ_y and ϕ_x show the rotations of the normal to the mid-plane about the y and x directions, respectively. The strain–displacement relations for this theory are:

$$\varepsilon_{xx} = \frac{\partial u}{\partial x} + z\frac{\partial \phi_x}{\partial x}, \tag{1.36}$$

$$\varepsilon_{yy} = \frac{\partial v}{\partial y} + z\frac{\partial \phi_y}{\partial y}, \tag{1.37}$$

$$\varepsilon_{xy} = \frac{\partial v}{\partial x} + \frac{\partial u}{\partial y} + z\left(\frac{\partial \phi_y}{\partial x} + \frac{\partial \phi_x}{\partial y}\right), \tag{1.38}$$

$$\varepsilon_{xz} = \phi_x + \frac{\partial w}{\partial x}, \tag{1.39}$$

$$\varepsilon_{zy} = \frac{\partial w}{\partial y} + \phi_y. \tag{1.40}$$

1.4.3 SINUSOIDAL SHEAR DEFORMATION PLATE THEORY

The sinusoidal shear deformation plate theory is an advanced method used to analyze the behavior of thin plates subjected to bending loads. Unlike classical plate theories, which assume constant shear deformation distributions through the plate thickness, the sinusoidal shear deformation theory incorporates a sinusoidal variation in shear strains across the thickness. This theory acknowledges the

non-uniform distribution of shear stresses and strains within the plate, particularly
in cases where the plate's cross-section is asymmetric or varies along its length.
By considering this sinusoidal variation, the theory provides a more accurate representation of shear deformations, especially in plates with thick or non-uniform
cross-sections.

One of the main advantages of the sinusoidal shear deformation theory is its ability to capture more complex deformation patterns, resulting in improved accuracy in
predicting plate behavior. Engineers can use this theory to obtain more precise calculations of deflections, stresses, and failure modes, particularly in structures subjected
to significant bending loads or with challenging geometries. In engineering practice, the sinusoidal shear deformation theory is applied in various fields, including
aerospace engineering, civil engineering, and mechanical engineering. Its advanced
capabilities make it a valuable tool for designing and analyzing a wide range of structures, from aircraft components to building facades. By considering the sinusoidal
variation in shear deformations, engineers can develop safer, more efficient designs
that meet the demands of modern engineering challenges. Based on this theory, we
have [1]:

$$u_x(x,y,z,t) = u(x,y,t) - z\frac{\partial w_b}{\partial x} - \underbrace{\left(z - \left(\frac{h}{\pi}\sin\frac{\pi z}{h}\right)\right)}_{f(z)}\frac{\partial w_s}{\partial x}, \tag{1.41}$$

$$u_y(x,y,z,t) = v(x,y,t) - z\frac{\partial w_b}{\partial y} - \underbrace{\left(z - \left(\frac{h}{\pi}\sin\frac{\pi z}{h}\right)\right)}_{f(z)}\frac{\partial w_s}{\partial y}, \tag{1.42}$$

$$u_z(x,y,z,t) = w_b(x,y,t) + w_s(x,y,t). \tag{1.43}$$

The strain–displacement relations for this theory are:

$$\varepsilon_{xx} = \frac{\partial u}{\partial x} - z\frac{\partial^2 w_b}{\partial x^2} - f\frac{\partial^2 w_s}{\partial x^2}, \tag{1.44}$$

$$\varepsilon_{yy} = \frac{\partial v}{\partial y} - z\frac{\partial^2 w_b}{\partial y^2} - f\frac{\partial^2 w_s}{\partial y^2}, \tag{1.45}$$

$$\varepsilon_{xy} = \frac{\partial u}{\partial y} + \frac{\partial v}{\partial x} - 2z\frac{\partial^2 w_b}{\partial x\partial y} - 2f\frac{\partial^2 w_s}{\partial x\partial y}, \tag{1.46}$$

$$\varepsilon_{yz} = g\frac{\partial w_s}{\partial y}, \tag{1.47}$$

$$\varepsilon_{xz} = g\frac{\partial W_s}{\partial x}. \tag{1.48}$$

1.5 SHELL THEORIES

1.5.1 CLASSICAL SHELL THEORY

In the classical shell theory or Kirchhoff-Love, a basic relation is expressed for the analysis of thin cylindrical shells subjected to various loads. This theory is based on the assumptions that shells are thin structures with negligible bending resistance in the thickness direction and that the middle surface of the shell remains undeformed and perpendicular to the reference plane. In classical shell theory, shells are treated as two-dimensional surfaces with three translational and three rotational degrees of freedom at each point. This allows for the derivation of governing equations that describe the deformation and stress distribution within the shell under different loading conditions.

One of the key features of classical shell theory is its ability to capture the essential characteristics of shell behavior, including membrane effects (tensile and compressive stresses resulting from in-plane loads) and bending effects (stresses resulting from out-of-plane deformations). By considering both membrane and bending effects, engineers can obtain accurate predictions of deflections, stresses, and failure modes within thin shell structures. Classical shell theory serves as a valuable tool for analyzing a wide range of shell structures, including pressure vessels, aircraft fuselages, and cylindrical tanks. It provides engineers with a simplified yet effective approach for designing and analyzing these structures, enabling them to optimize designs, ensure structural integrity, and meet performance requirements.

Despite its simplicity, classical shell theory has limitations, particularly for shells with complex geometries or subjected to large deformations. In such cases, more advanced shell theories, such as the Reissner–Mindlin theory or higher-order shell theories, may be necessary to provide more accurate predictions. Nonetheless, for many engineering applications, classical shell theory remains a practical and widely used tool for analyzing thin shell structures and informing design decisions. Based on this theory, we have [1, 2]:

$$u_x(x,\theta,z,t) = u(x,\theta,t) - z\frac{\partial w(x,\theta,t)}{\partial x}, \tag{1.49}$$

$$u_\theta(x,\theta,z,t) = v(x,\theta,t) - \frac{z}{R}\frac{\partial w(x,\theta,t)}{\partial \theta}, \tag{1.50}$$

$$u_z(x,\theta,z,t) = w(x,\theta,t). \tag{1.51}$$

The strain–displacement relations for this theory are:

$$\varepsilon_{xx} = \frac{\partial u}{\partial x} - z\frac{\partial^2 w}{\partial x^2}, \tag{1.52}$$

$$\varepsilon_{\theta\theta} = \frac{\partial v}{R\partial\theta} + \frac{w}{R} - \frac{z}{R^2}\frac{\partial^2 w}{\partial \theta^2}, \tag{1.53}$$

$$\varepsilon_{x\theta} = \frac{1}{2}\left(\frac{\partial v}{R\partial \theta} + \frac{\partial v}{\partial x}\right) - z\frac{\partial^2 w}{R\partial x\partial \theta}. \tag{1.54}$$

1.5.2 First-Order Shear Deformation Shell Theory

The first-order shear deformation shell theory is an advancement of classical shell theory, offering a more accurate representation of thin shell behavior. Unlike classical shell theory, which assumes that the middle surface of the shell remains undeformed and perpendicular to the reference plane, the first-order shear deformation theory considers the effects of shear deformation within the shell thickness. In this theory, the displacement field is allowed to vary linearly through the thickness of the shell, capturing the shearing effects that occur as the shell deforms. By incorporating this shear deformation, the theory provides a more accurate representation of the shell's behavior, particularly for shells with moderate thicknesses or subjected to significant bending loads.

One of the main advantages of the first-order shear deformation shell theory is its ability to predict more realistic deflections and stress distributions within the shell. By considering the variation in displacement through the thickness, engineers can obtain more accurate results for shells with non-uniform cross-sections or complex loading conditions. In engineering practice, the first-order shear deformation shell theory finds widespread use in the analysis and design of thin and moderately thick shell structures. It serves as a valuable tool for predicting deflections, stresses, and buckling modes, particularly in aerospace, civil, and mechanical engineering applications.

Despite its advancements, the first-order shear deformation shell theory still has limitations, particularly for shells with very thick cross-sections or highly nonlinear material behavior. In such cases, higher-order shell theories or finite element analysis may be necessary to provide more accurate predictions. Nonetheless, for many engineering applications, the first-order shear deformation shell theory remains a practical and effective tool for analyzing shell structures and informing design decisions. Based on this theory, we have [3]:

$$u_x(x,\theta,z,t) = u(x,\theta,t) + z\psi_x(x,\theta,t), \tag{1.55}$$

$$u_\theta(x,\theta,z,t) = v(x,\theta,t) + z\psi_\theta(x,\theta,t), \tag{1.56}$$

$$u_z(x,\theta,z,t) = w(x,\theta,t). \tag{1.57}$$

The strain–displacement relations for this theory are:

$$\varepsilon_{xx} = \frac{\partial u}{\partial x} + z\frac{\partial \psi_x}{\partial x}, \tag{1.58}$$

$$\varepsilon_{\theta\theta} = \frac{\partial v}{R\partial\theta} + \frac{w}{R} + z\frac{\partial\psi_\theta}{R\partial\theta}, \tag{1.59}$$

$$\gamma_{x\theta} = \frac{\partial v}{\partial x} + \frac{\partial u}{R\partial\theta} + z\left(\frac{\partial\psi_x}{R\partial\theta} + \frac{\partial\psi_\theta}{\partial x}\right), \tag{1.60}$$

$$\varepsilon_{xz} = \frac{\partial w}{\partial x} + \psi_x, \tag{1.61}$$

$$\varepsilon_{\theta z} = \frac{\partial w}{R\partial\theta} - \frac{v}{R} + \psi_\theta. \tag{1.62}$$

1.5.3 SINUSOIDAL SHEAR DEFORMATION SHELL THEORY

The sinusoidal shear deformation shell theory is an advanced method for analyzing the behavior of thin shells under various loads. Unlike traditional shell theories that assume a uniform shear deformation across the shell's thickness, this theory incorporates a sinusoidal variation in shear strains through the thickness. This approach recognizes the non-uniform distribution of shear stresses and strains within the shell, especially when the shell's cross-section is asymmetrical or varies along its length. By including this sinusoidal variation, the theory provides a more precise representation of shear deformations, making it particularly effective for shells with thick or irregular cross-sections.

One of the main advantages of the sinusoidal shear deformation theory is its ability to capture more complex deformation patterns, leading to improved accuracy in predicting plate behavior. Engineers can use this theory to obtain more precise calculations of deflections, stresses, and failure modes, particularly in structures subjected to significant bending loads or with challenging geometries. In engineering practice, the sinusoidal shear deformation theory finds applications in various fields, including aerospace engineering, civil engineering, and mechanical engineering. Its advanced capabilities make it a valuable tool for designing and analyzing a wide range of structures, from aircraft components to building facades. By considering the sinusoidal variation in shear deformations, engineers can develop safer, more efficient designs that meet the demands of modern engineering challenges. Based on this theory, we have [4]:

$$u_x(x,\theta,z,t) = u(x,\theta,t) - z\frac{\partial w_b}{\partial x} - \underbrace{\left(z - \left(\frac{h}{\pi}\sin\frac{\pi z}{h}\right)\right)}_{f(z)}\frac{\partial w_s}{\partial x}, \tag{1.63}$$

$$u_\theta(x,\theta,z,t) = v(x,\theta,t) - z\frac{\partial w_b}{R\partial\theta} - \underbrace{\left(z - \left(\frac{h}{\pi}\sin\frac{\pi z}{h}\right)\right)}_{f(z)}\frac{\partial w_s}{R\partial\theta}, \tag{1.64}$$

$$u_z(x,\theta,z,t) = w_b(x,\theta,t) + w_s(x,\theta,t). \tag{1.65}$$

The strain–displacement relations for this theory are:

$$\varepsilon_{xx} = \frac{\partial u}{\partial x} - z\frac{\partial^2 w_b}{\partial x^2} - f\frac{\partial^2 w_s}{\partial x^2}, \tag{1.66}$$

$$\varepsilon_{\theta\theta} = \frac{\partial v}{R\partial\theta} - z\frac{\partial^2 w_b}{R^2\partial\theta^2} - f\frac{\partial^2 w_s}{R^2\partial\theta^2}, \tag{1.67}$$

$$\varepsilon_{x\theta} = \frac{\partial u}{R\partial\theta} + \frac{\partial v}{\partial x} - 2z\frac{\partial^2 w_b}{R\partial x\partial\theta} - 2f\frac{\partial^2 w_s}{R\partial x\partial\theta}, \tag{1.68}$$

$$\varepsilon_{\theta z} = g\frac{\partial w_s}{R\partial\theta}, \tag{1.69}$$

$$\varepsilon_{xz} = g\frac{\partial W_s}{\partial x}. \tag{1.70}$$

REFERENCES

1. Reddy JN. *Mechanics of laminated composite plates and shells: Theory and analysis.* 2nd edition. CRC Press, Boca Raton, USA, 2002.
2. Reddy JN. A simple higher order theory for laminated composite plates. *J. Appl. Mech.* 1984;51:745–752.
3. Brush O, Almorth B. *Buckling of bars, plates and shells.* Mc-Graw Hill, New York, USA, 1975.
4. Thai HT, Vo TP. A new sinusoidal shear deformation theory for bending buckling and vibration of functionally graded plates. *Appl. Math. Model.* 2013;37(5):3269–3281.

2 Numerical Methods

2.1 INTRODUCTION

Numerical methods are indispensable tools in engineering and scientific research, providing efficient and accurate solutions to complex problems that defy analytical approaches. In this comprehensive review, we explore four prominent numerical techniques: the differential quadrature method (DQM), differential cubature method (DCM), Galerkin method, and Navier method. These methods have wide-ranging applications across various fields, including structural engineering, fluid dynamics, electromagnetics, and heat transfer. By delving into their principles, advantages, limitations, and applications, this review aims to offer a thorough understanding of these numerical techniques.

- Differential Quadrature Method (DQM): The DQM is a powerful numerical approach for solving differential equations and boundary value problems. Rooted in numerical analysis, the DQM discretizes derivatives of a function at discrete points using weighted sums of function values. Unlike finite difference methods, the DQM offers higher accuracy and convergence rates, making it suitable for problems with intricate geometries and irregular boundaries. Its versatility allows it to tackle various types of differential equations, including ordinary and partial differential equations, as well as integro-differential equations.
- Differential Cubature Method (DCM): The DCM extends the principles of the DQM to multidimensional spaces. While the DQM discretizes derivatives along individual coordinate axes, the DCM employs cubature rules to approximate derivatives in multiple directions. This capability enables the DCM to handle problems with higher dimensions efficiently, such as those encountered in computational fluid dynamics and structural mechanics. By leveraging higher-order cubature rules, the DCM achieves superior accuracy and convergence rates compared to traditional finite difference methods, making it well-suited for high-dimensional and complex problems.
- Galerkin Method: The Galerkin method is a widely utilized numerical technique for solving differential equations by approximating the solution using a finite-dimensional subspace of the function space. By selecting suitable basis functions, such as polynomials or trigonometric functions, the Galerkin method minimizes the residual error in the differential equation over the chosen subspace. This yields a set of algebraic equations that can be solved to obtain an approximate solution. The Galerkin method is particularly effective for linear partial differential equations, boundary value problems, and eigenvalue problems.

DOI: 10.1201/9781003510710-2

- Navier Method: The Navier method, named after the pioneering physicist Claude-Louis Navier, is a classical approach for solving fluid flow problems. It approximates the fluid velocity field and pressure distribution based on the Navier–Stokes equations, which govern the motion of viscous fluids. By discretizing the equations in both space and time domains, the Navier method obtains a numerical solution. While relatively simple, the Navier method is widely used in computational fluid dynamics due to its computational efficiency and ease of implementation, particularly for problems involving laminar flow and low Reynolds numbers.

In conclusion, the DQM, DCM, Galerkin method, and Navier method are fundamental numerical techniques with diverse applications in engineering and scientific research. By elucidating the theoretical foundations, computational procedures, and practical considerations associated with each method, this chapter aims to provide valuable insights into their strengths, limitations, and future research directions. Through a comprehensive examination of these numerical methods, researchers and practitioners can leverage their capabilities to address a wide array of complex problems across various disciplines.

2.2 DIFFERENTIAL QUADRATURE METHOD

The DQM is a powerful numerical technique widely employed for solving differential equations and boundary value problems in engineering and scientific research. The DQM has gained popularity due to its high accuracy, flexibility, and efficiency in handling complex problems with irregular geometries. This concise introduction provides an overview of the fundamental principles, computational procedure, and applications of the DQM.

- Principles of the DQM: At its core, the DQM discretizes derivatives of a function by expressing them as weighted sums of function values at selected grid points. Unlike finite difference methods, which approximate derivatives using Taylor series expansions, the DQM employs quadrature rules to accurately estimate derivatives at discrete locations. By judiciously choosing the spacing and distribution of grid points, the DQM achieves high-order accuracy, enabling precise solutions even for nonlinear and multidimensional problems.
- Computational Procedure: The computational procedure of the differential quadrature method involves several key steps:
  - Grid Generation: The domain of the problem is discretized into a set of grid points, typically arranged in a regular or non-uniform manner. The number and distribution of grid points depend on the complexity of the problem and the desired accuracy of the solution.
  - Approximation of Derivatives: At each grid point, derivatives of the function are approximated using quadrature rules. These rules involve weighted combinations of function values at neighboring grid points, with weights determined based on the chosen quadrature scheme.

- Construction of Difference Equations: The differential equation governing the problem is discretized using the approximated derivatives, resulting in a system of algebraic equations. These equations are formulated by enforcing the boundary conditions and constraints imposed by the problem.
- Solution of Algebraic Equations: The system of algebraic equations obtained from the discretization process is solved using numerical methods such as Gaussian elimination, lower–upper (LU) decomposition, or iterative solvers. The solution provides an approximation to the desired solution of the original differential equation.

- Applications of the DQM: The DQM finds widespread application in various fields of engineering and science, including structural mechanics, fluid dynamics, heat transfer, electromagnetics, and acoustics. Some common applications of DQM include:
  - Structural Analysis: The DQM is used to analyze the static and dynamic behavior of structures subjected to mechanical loads, such as beams, plates, shells, and composite materials.
  - Heat Transfer: The DQM is applied to model heat conduction and convection in thermal systems, including heat exchangers, fins, and electronic devices.
  - Fluid Dynamics: The DQM is employed to simulate fluid flow phenomena, such as laminar and turbulent flow, boundary layer separation, and vortex shedding.
  - Electromagnetics: The DQM is utilized to solve Maxwell's equations for electromagnetic wave propagation, antenna design, and microwave devices.

- Advantages:
  - High Accuracy: The DQM is known for its high accuracy in approximating derivatives and solving differential equations. By employing interpolation polynomials to approximate derivatives, The DQM can achieve very accurate solutions, even with relatively few discretization points.
  - Versatility: The DQM can be applied to a wide range of problems, including ordinary and partial differential equations in various domains such as mechanics, heat transfer, electromagnetics, and fluid dynamics. Its versatility makes it a valuable tool for engineers and scientists working in diverse fields.
  - Efficiency: The DQM often requires fewer grid points compared to finite difference methods, leading to more efficient computations. This efficiency is particularly advantageous for problems with high computational costs or large computational domains.
  - Stability: The DQM is numerically stable for many types of differential equations, even those with stiff or oscillatory solutions. This stability ensures that the method produces reliable results without numerical instabilities or oscillations.

- Limitations:

    - Boundary Conditions: The DQM can encounter challenges when dealing with certain types of boundary conditions, particularly those involving non-standard or complex geometries. In such cases, specialized techniques or modifications may be necessary to accurately enforce boundary conditions.
    - Computational Cost: While the DQM is generally more efficient than finite difference methods, it can still be computationally expensive for problems with large numbers of grid points or complex discretization schemes. As a result, the method may not be well-suited for all computational tasks, especially those with tight computational budgets.
    - Discretization Errors: Like any numerical method, the DQM is subject to discretization errors, particularly when using coarse grids or low-order interpolation schemes. These errors can lead to inaccuracies in the solution, especially in regions with steep gradients or rapid variations.
    - Implementation Complexity: Implementing the DQM for complex problems may require a thorough understanding of numerical analysis and programming skills. Developing efficient and accurate numerical schemes often requires the careful consideration of discretization strategies, interpolation schemes, and computational algorithms.

The DQM is a versatile and efficient numerical technique for solving differential equations and boundary value problems encountered in engineering and scientific research. Its ability to achieve high accuracy and flexibility makes it an invaluable tool for analyzing complex systems and phenomena. As computational resources continue to advance, the DQM remains a key methodological approach for tackling challenging problems across various disciplines.

In this method, the differential equations are changed into a first-order algebraic equation by employing appropriate weighting coefficients because weighting coefficients do not relate to any special problem and only depend on the grid spacing. In other words, the partial derivatives of a function (say w here) are approximated with respect to specific variables (say x and y) at a discontinuous point in a defined domain ($0 < x < L_x$ and $0 < y < L_y$) as a set of linear weighting coefficients and the amount represented by the function itself at that point and other points throughout the domain. The approximation of the nth and m^{th} derivatives function with respect to x and y, respectively, may be expressed in general form as [1]:

$$f_x^{(n)}(x_i,y_i) = \sum_{k=1}^{N_x} A^{(n)}{}_{ik} f(x_k,y_j),$$

$$f_y^{(m)}(x_i,y_i) = \sum_{l=1}^{N_y} B^{(m)}{}_{jl} f(x_i,y_l), \tag{2.1}$$

$$f_{xy}^{(n+m)}(x_i,y_i) = \sum_{k=1}^{N_x} \sum_{l=1}^{N_y} A^{(n)}{}_{ik} B^{(m)}{}_{jl} f(x_k,y_l),$$

where N_x and N_y denote the number of points in the x and y directions, $f(x, y)$ is the function, and A_{ik}, B_{jl} are the weighting coefficients defined as:

$$A^{(1)}_{ij} = \frac{M(x_i)}{(x_i - x_j)M(x_j)},$$

$$B^{(1)}_{ij} = \frac{P(y_i)}{(y_i - y_j)M(y_j)},$$

(2.2)

where M and P are Lagrangian operators defined as:

$$M(x_i) = \prod_{j=1}^{N_x}(x_i - x_j), \quad i \neq j$$

$$P(y_i) = \prod_{j=1}^{N_y}(y_i - y_j), \quad i \neq j.$$

(2.3)

The weighting coefficients for the second, third, and fourth derivatives are determined via matrix multiplication:

$$A^{(2)}_{ij} = \sum_{k=1}^{N_x} A^{(1)}_{ik} A^{(1)}_{kj}, \qquad A^{(3)}_{ij} = \sum_{k=1}^{N_x} A^{(2)}_{ik} A^{(1)}_{kj},$$

$$A^{(4)}_{ij} = \sum_{k=1}^{N_x} A^{(3)}_{ik} A^{(1)}_{kj}, \qquad i, j = 1, 2, ..., N_x,$$

$$B^{(2)}_{ij} = \sum_{k=1}^{N_y} B^{(1)}_{ik} B^{(1)}_{kj}, \qquad B^{(3)}_{ij} = \sum_{k=1}^{N_y} B^{(2)}_{ik} B^{(1)}_{kj},$$

$$B^{(4)}_{ij} = \sum_{k=1}^{N_y} B^{(3)}_{ik} B^{(1)}_{kj}, \qquad i, j = 1, 2, ..., N_y.$$

(2.4)

Using the following rule, the distribution of grid points in domain is calculated as:

$$x_i = \frac{L_x}{2}\left[1 - \cos\left(\frac{\pi i}{N_x}\right)\right],$$

$$y_j = \frac{L_y}{2}\left[1 - \cos\left(\frac{\pi j}{N_y}\right)\right].$$

(2.5)

From this method, an example in MatLab is presented in Appendix A.

2.3 DIFFERENTIAL CUBATURE METHOD

The differential cubature method (DCM) is a numerical technique used for solving differential equations, particularly those involving partial derivatives. Unlike traditional methods such as finite differences or finite elements, which approximate derivatives using discretization schemes, the DCM relies on integration formulas to directly evaluate the differential operators in the differential equations.

In the differential cubature method, the domain of interest is discretized into a set of cubature points, where the differential equations are evaluated. Integration formulas, such as Gaussian quadrature or cubature rules, are then applied to numerically compute the integrals involving the derivatives. By directly evaluating the derivatives at the cubature points, DCM bypasses the need for explicit derivative approximations, leading to more accurate and efficient solutions. One of the key advantages of the differential cubature method is its ability to achieve high accuracy with relatively few discretization points. By leveraging integration formulas, the DCM can accurately capture the behavior of the differential equations, even in regions with steep gradients or rapid variations. This accuracy is particularly advantageous for problems with complex geometries or nonlinear behavior, where traditional discretization methods may struggle to provide accurate solutions.

Additionally, the differential cubature method offers flexibility in choosing the distribution of cubature points and the integration rules used for numerical integration. This flexibility allows engineers and scientists to tailor the discretization scheme to the specific characteristics of the problem, optimizing accuracy and computational efficiency. Despite its advantages, the DCM may encounter challenges when dealing with certain types of differential equations or boundary conditions. In some cases, specialized techniques or modifications may be necessary to accurately handle non-standard or complex problems. Additionally, the computational cost of the DCM can be higher than that of simpler discretization methods, especially for problems with large numbers of cubature points or high-dimensional domains.

Overall, the differential cubature method is a powerful numerical technique for solving differential equations, offering high accuracy and flexibility in a wide range of applications. By directly evaluating derivatives through integration formulas, the DCM provides a robust framework for tackling complex engineering and scientific problems with confidence.

- Principles of the DCM: The DCM is founded on the principle of directly evaluating the derivatives of differential equations using integration formulas. Instead of discretizing the domain with mesh points and approximating derivatives, the DCM employs cubature points distributed within the domain. At these cubature points, integration formulas, such as Gaussian quadrature or other cubature rules, are applied to accurately compute the integrals involving the derivatives. This direct approach bypasses the need for explicit derivative approximations, leading to more accurate and efficient solutions.
- Computational Procedure:
  - Cubature Point Selection: Cubature points are distributed within the domain of interest.
  - Numerical Integration: Integration formulas are applied to the cubature points to compute the integrals involving the derivatives.

- System Assembly: The system of equations is assembled based on the discretization and the differential equations.
- Solution: The resulting system of equations is solved to obtain the solution.

- Applications of the DCM:
  - The DCM finds applications across various fields, including:
  - Partial Differential Equations (PDEs): The DCM is effective for solving PDEs governing phenomena like heat transfer, fluid dynamics, and structural mechanics.
  - Inverse Problems: The DCM can be employed in inverse problems, such as parameter estimation and image reconstruction.
  - Optimization: The DCM can assist in optimization tasks by providing accurate gradients required for optimization algorithms.

- Advantages of the DCM:
  - High Accuracy: The DCM achieves high accuracy by directly evaluating derivatives, even in regions with steep gradients or rapid variations. Flexibility: The DCM offers flexibility in choosing cubature points and integration rules tailored to specific problem characteristics.
  - Efficiency: Despite its accuracy, the DCM can be computationally efficient, particularly for problems with complex geometries or nonlinear behavior.

- Limitations of the DCM:

  - Computational Cost: The DCM can be computationally expensive, especially for problems with large numbers of cubature points or high-dimensional domains.
  - Challenges with Boundary Conditions: The DCM may encounter challenges in handling certain boundary conditions, particularly those involving non-standard or complex geometries.
  - Implementation Complexity: Implementing the DCM for complex problems requires a thorough understanding of numerical methods and programming skills.

In summary, the DCM offers high accuracy and flexibility, making it a valuable tool for solving differential equations across diverse engineering and scientific applications. However, its computational cost and challenges with certain boundary conditions should be carefully considered when applying the method to real-world problems.

The DCM is a powerful numerical technique employed in various engineering and scientific applications. It operates by representing the value of a function at a discrete point within the solution domain through a weighted linear combination of discrete function values selected across the entire problem domain. Particularly for two-dimensional problems, where an array of arbitrarily located grid points is assumed, DCM offers a precise cubature approximation at the ith discrete point. This method is instrumental in accurately solving differential equations and simulating complex phenomena, providing invaluable insights into diverse real-world problems. For a 2D problem, we have [2]:

$$\Im F\left(x,y\right)_i \approx \sum_{j=1}^{N} D_{ij} F(x_j, y_j), \tag{2.6}$$

where $\Im$ is the operator, N is the grid point number and D_{ij} is the weighting cubature factor, which may be calculated by:

$$\Im\left\{x^{a-b}y^b\right\}_i = \sum_{j=1}^{N} D_{ij} F(x_j^{a-b}\theta_j^b),\ b = 0,1,2,...,\nu,\ a = 0,1,2,...,N-1,\ i = 1,2,...,N, \tag{2.7}$$

where

$$\left[x_j^{a-b}y_j^b\right] \begin{bmatrix} D_{i1} \\ D_{i2} \\ \cdot \\ \cdot \\ \cdot \\ D_{in} \end{bmatrix} = \left[\Im\left\{x^{a-b}y^b\right\}_i\right]. \tag{2.8}$$

2.4 GALERKIN METHOD

The Galerkin method is a powerful numerical technique used for solving differential equations and boundary value problems in various fields of engineering, physics, and applied mathematics. Named after the Russian mathematician Boris Galerkin, who introduced the method in the early 20th century, the Galerkin method offers a systematic approach to approximating the solution of differential equations by projecting the problem onto a finite-dimensional subspace.

At its core, the Galerkin method seeks an approximate solution to a given differential equation by formulating it as a variational problem. The key idea is to seek a solution within a finite-dimensional function space, typically spanned by a set of basis functions. These basis functions are chosen to reflect the underlying physics of the problem and may include polynomials, trigonometric functions, or other suitable functions. The Galerkin method works by minimizing the residual, or error, between the true solution of the differential equation and its approximation within the chosen function space. This is achieved by forcing the residual to be orthogonal to the subspace spanned by the basis functions. By satisfying this orthogonality condition, the Galerkin method ensures that the approximation converges to the true solution as the number of basis functions increases.

The computational procedure of the Galerkin method involves several key steps:

- Formulation of the Variational Problem: The differential equation governing the problem is recast into a variational form using concepts from calculus of variations. This involves multiplying the equation by a test function and integrating over the domain of interest.
- Selection of Basis Functions: A set of basis functions is chosen to span the function space in which the approximate solution will be sought. The choice of basis functions depends on the problem at hand and may involve polynomials, piecewise functions, or other suitable functions.
- Discretization of the Domain: The domain of the problem is discretized into a finite number of elements or nodes. This may involve subdividing the domain into finite elements in the case of partial differential equations or using a grid-based approach for ordinary differential equations.
- Assembly of the System of Equations: The variational problem leads to a system of algebraic equations, known as the Galerkin equations, which must be solved to obtain the approximate solution. This system is typically obtained by discretizing the variational problem using the chosen basis functions and performing numerical integration over the domain.

The Galerkin method finds widespread application in various areas of engineering and science, including structural mechanics, fluid dynamics, heat transfer, electromagnetics, and quantum mechanics. Some common applications of the Galerkin method include:

- Finite Element Analysis: The Galerkin method serves as the foundation for the finite element method, a widely used technique for solving partial differential equations in structural and solid mechanics.
- Computational Fluid Dynamics: The Galerkin method is employed to discretize and solve the Navier–Stokes equations for simulating fluid flow phenomena in aerospace, automotive, and environmental engineering.
- Heat Transfer Analysis: The Galerkin method is applied to model heat conduction and convection in thermal systems, including heat exchangers, electronic devices, and building energy systems.

The Galerkin method is a versatile and effective numerical technique for solving differential equations and boundary value problems encountered in engineering and scientific research. Its ability to provide accurate solutions across a wide range of applications makes it an invaluable tool for analyzing complex systems and phenomena. As computational resources continue to advance, the Galerkin method remains a cornerstone of numerical analysis and simulation in diverse fields of study.

The basic relation for this method is [3]:

$$Y(x,y,t) = \Psi(x,y)\Upsilon(t), \tag{2.9}$$

where $\Psi(x,y)$ and $\Upsilon(t)$ are location- and time-dependent variables, respectively. By substituting this relation into motion equations, the algebraic relations may be derived. The location-dependent variables are chosen based on various boundary conditions.

2.5 NAVIER METHOD

The Navier method, named after the French physicist and engineer Claude-Louis Navier, stands as a cornerstone in the realm of structural engineering and applied mechanics. Developed in the early 19th century, this mathematical approach is pivotal in understanding the deformation and stress distribution within solid bodies subjected to external loads. It finds extensive application across various fields, including civil engineering, materials science, and mechanical engineering, serving as a fundamental tool for analyzing the response of engineering structures under different loading conditions.

At its essence, the Navier method relies on the principles of continuum mechanics and the theory of elasticity to describe the mechanical behavior of deformable solids. The method assumes that the material behaves linearly elastic, meaning it deforms proportionally to the applied load within the elastic limit. The governing equations of the Navier method stem from the equilibrium equations and the constitutive relation between stress and strain, commonly known as Hooke's law.

The computational procedure of the Navier method involves several key steps. First, the equilibrium equations governing the static behavior of the structure are formulated based on Newton's laws of motion and the principles of equilibrium. These equations relate the external loads applied to the structure to the internal stresses and deformations within the material. Second, the constitutive relation between stress and strain is established based on Hooke's law, defining the linear elastic behavior of the material. Third, boundary conditions are applied to the structure to define the constraints and loading conditions at its boundaries. These conditions may include fixed displacements, prescribed loads, or frictional contact with other surfaces. Lastly, the equilibrium equations, together with the constitutive relation and boundary conditions, form a system of partial differential equations that describe the behavior of the structure. These equations are solved using numerical techniques such as finite difference, finite element, or boundary element methods to obtain the displacement field and stress distribution within the structure.

The Navier method finds a wide range of applications in structural engineering, from analyzing the stress and deformation in beams, plates, and shells to determining the mechanical properties of materials. It is used in structural analysis to predict the behavior of bridges, buildings, and mechanical components under static loads. Moreover, it serves as a valuable tool for material characterization, aiding in the determination of Young's modulus, Poisson's ratio, and shear modulus through experimental testing and numerical simulation. Additionally, engineers utilize the Navier method for design optimization, enabling the optimization of structural configurations and materials for improved performance and efficiency under various loading conditions. In conclusion, the Navier method remains an indispensable tool in structural engineering and applied mechanics, offering valuable insights into the behavior of deformable solids under external loads. Its versatility and accuracy make it a cornerstone of engineering analysis, supporting the development of innovative and efficient engineering solutions. As computational methods continue to advance, the Navier method stands poised to play a central role in addressing the challenges of modern engineering design and analysis.

This method is based on the simply supported boundary conditions for which $\Psi(x, y)$ may be written as [4]:

$$u(x,y) = \sum_{m=1}^{\infty}\sum_{n=1}^{\infty} u_{mn} \cos\left(\frac{m\pi x}{a}\right)\sin\left(\frac{n\pi y}{b}\right), \qquad (2.10)$$

$$v(x,y) = \sum_{m=1}^{\infty}\sum_{n=1}^{\infty} v_{mn} \sin\left(\frac{m\pi x}{a}\right)\cos\left(\frac{n\pi y}{b}\right), \qquad (2.11)$$

$$w(x,y) = \sum_{m=1}^{\infty}\sum_{n=1}^{\infty} w_{mn} \sin\left(\frac{m\pi x}{a}\right)\sin\left(\frac{n\pi y}{b}\right), \qquad (2.12)$$

$$\varphi_x(x,y) = \sum_{m=1}^{\infty}\sum_{n=1}^{\infty} \varphi_{xmn} \cos\left(\frac{m\pi x}{a}\right)\sin\left(\frac{n\pi y}{b}\right), \qquad (2.13)$$

$$\varphi_y(x,y) = \sum_{m=1}^{\infty}\sum_{n=1}^{\infty} \varphi_{ymn} \sin\left(\frac{m\pi x}{a}\right)\cos\left(\frac{n\pi y}{b}\right). \qquad (2.14)$$

From this method, an example in MatLab is presented in Appendix B.

REFERENCES

1. Kolahchi R, Safari M, Esmailpour M. Dynamic stability analysis of temperature-dependent functionally graded CNT-reinforced visco-plates resting on orthotropic elastomeric medium. *Compos. Struct.* 2016;150:255–265.
2. Kolahchi R, Hosseini H, Esmailpour M. Differential cubature and quadrature-Bolotin methods for dynamic stability of embedded piezoelectric nanoplates based on visco-nonlocalpiezoelasticity theories. *Compos. Struct.* 2016;157:174–186.
3. Brenner S, Scott RL. *The mathematical theory of finite element methods.* 2nd edition. New York: Springer, 2005. ISBN 0-387-95451-1.
4. Akhavan H, Hosseini Hashemi Sh, Rokni Damavandi Taher H, Alibeigloo A, Vahabi Sh. Exact solutions for rectangular Mindlin plates underin-plane loads resting on Pasternak elastic foundation. Part I: Buckling analysis. *Comput. Mat. Sci.* 2009;44(3):968–978.

3 Buckling of Nanoparticle-Reinforced Beams

3.1 INTRODUCTION

Extensive empirical research has been conducted to model various structures, typically categorized into two main approaches: atomic modeling and continuum mechanics. Atomic modeling techniques encompass molecular dynamics, molecular dynamics with strong bonds, and the theory of density. However, atomic modeling involves intricate calculations for systems with numerous atoms, and its practical applications are often limited. Conversely, the development of continuum mechanics has enabled the overcoming of the limitations associated with atomic modeling. Today, continuum mechanics models are widely employed for structural modeling. A comparison of atomic and continuum mechanics modeling results reveals that continuum mechanics provide broadly acceptable predictions for the dynamic and static behavior of systems. Consequently, most researchers investigating the dynamic and static behavior of structures rely on continuum mechanics.

The referenced papers cover a diverse range of topics in structural engineering and materials science. Formica et al. [1] delved into the vibrations of carbon nanotube-reinforced composites, shedding light on their dynamic behavior. Ghorbanpour Arani et al. [2] contributed to the field with a study on buckling analysis and smart control of structures using polyvinylidene fluoride (PVDF)nanoplates, while their next work explored nonlinear surface and nonlocal piezoelasticity theories for the vibration analysis of boron nitride sheets [3]. Additionally, Ghorbanpour Arani et al. [4] conducted a static stress analysis of carbon nanotube-reinforced composite cylinders under various loading conditions, and Henkhaus et al. [5] investigated the axial failure of reinforced concrete columns damaged by shear reversals. Jafarian Arani and Kolahchi [6] focused on the buckling analysis of embedded concrete columns armed with carbon nanotubes, whereas Kadoli and Ganesan [7] explored the free vibration and buckling analysis of composite cylindrical shells conveying hot fluid. Furthermore, Kolahchi et al. [8] presented a nonlocal nonlinear analysis for buckling in embedded micro plates subjected to a magnetic field, and their subsequent work addressed the dynamic stability analysis of temperature-dependent functionally graded materials [9]. Moreover, their study on the dynamic stability of embedded piezoelectric nanoplates contributes to understanding the behavior of piezoelectric materials under viscoelastic conditions [10]. Lastly, Karaca and Türkeli [11] investigated the slenderness effect on the wind response of industrial reinforced concrete chimneys, crucial for optimizing the design and stability of tall structures.

DOI: 10.1201/9781003510710-3

In recent years, there has been a growing interest in the mechanical behavior of various composite structures, especially those reinforced with nanomaterials. Researchers have extensively investigated the dynamic instability, postbuckling behavior, vibration, and buckling of such structures using advanced analytical and computational methods. Kolahchi et al. [12] studied the dynamic instability of single-walled carbon nanotubes utilizing a size-dependent sinusoidal beam model. Liew et al. [13] explored the postbuckling behavior of carbon nanotube-reinforced functionally graded cylindrical panels under axial compression, employing a meshless approach. Furthermore, Matsuna [14] delved into the vibration and buckling characteristics of cross-ply laminated composite circular cylindrical shells based on a global higher-order theory. Mirza and Skrabek [15] investigated the reliability of short composite beam–column strength interaction, shedding light on crucial structural aspects. In the realm of composite materials, Tan and Tong [16] developed microelectromechanical models specifically tailored for piezoelectric-fiber-reinforced composite materials. Additionally, Thai and Vo [17, 18] proposed both nonlocal sinusoidal shear deformation and nonlocal beam theories, offering insights into the bending, buckling, and vibration behaviors of nanobeams. Solhjoo and Vakis [19] conducted a comparison study between molecular dynamics simulations and continuum mechanics models for single asperity nanocontacts, providing valuable insights into nanoscale contact mechanics. Seo et al. [20] analyzed the frequency response of cylindrical shells conveying fluid using the finite element method, contributing to the understanding of fluid–structure interaction phenomena. Moreover, Wuite and Adali [21] investigated the deflection and stress behavior of nanocomposite-reinforced beams through a multiscale analysis, highlighting the importance of considering multiple length scales in modeling such structures. Additionally, Zamanian et al. [22] examined the agglomeration effects on the buckling behavior of embedded concrete columns reinforced with SiO_2 nanoparticles, providing insights into the structural performance of nanocomposite materials in practical applications.

No existing research has explored the impact of ZnO nanoparticles and an electric field on the buckling behavior of concrete columns. Therefore, this chapter introduces the concept of smart buckling in concrete columns reinforced with ZnO nanoparticles. Additionally, the presence of ZnO nanoparticles imparts piezoelectric properties to the structure, making it responsive to an electric field regardless of temperature variations. To analyze this phenomenon, a microelectromechanical model is employed to calculate the material properties of the structure. The concrete column is mathematically simulated using the sinusoidal shear deformation theory (SSDT) approach, and the differential quadrature method (DQM) is utilized to determine the buckling load of the structure. The study investigates the influence of various factors, including the volume percentage of ZnO nanoparticles, external voltage, boundary conditions, and foundation characteristics, on the buckling load of the system.

3.2 MOTION EQUATIONS

Figure 3.1 depicts a concrete column reinforced with ZnO nanoparticles. The column is situated within a foundation that is simulated using vertical spring and shear constants.

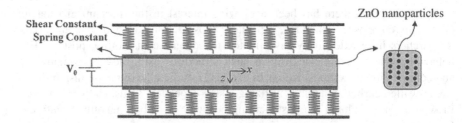

FIGURE 3.1 Diagram illustrating a concrete column reinforced with ZnO nanoparticles.

The beam is modeled with sinusoidal shear deformation beam theory. The displacement of this theory is presented in section 1.3.3 of Chapter 1. When piezoelectric materials experience mechanical stress, they produce an electric field within the material. Conversely, applying an electric field induces mechanical strain in the material. In such materials, the interaction between stress (σ) and strain (ε) from the mechanical perspective, alongside electrical displacement (D) and electric field (E) from the electrostatic viewpoint, can be coupled, as follows [2, 3]:

$$
\begin{bmatrix} \sigma_{xx} \\ \sigma_{yy} \\ \sigma_{zz} \\ \tau_{yz} \\ \tau_{xz} \\ \tau_{xy} \end{bmatrix} = \begin{bmatrix} C_{11} & C_{12} & C_{13} & 0 & 0 & 0 \\ C_{12} & C_{22} & C_{23} & 0 & 0 & 0 \\ C_{13} & C_{23} & C_{33} & 0 & 0 & 0 \\ 0 & 0 & 0 & C_{44} & 0 & 0 \\ 0 & 0 & 0 & 0 & C_{55} & 0 \\ 0 & 0 & 0 & 0 & 0 & C_{66} \end{bmatrix} \begin{bmatrix} \varepsilon_{xx} - \alpha_{xx}T \\ \varepsilon_{yy} - \alpha_{yy}T \\ \varepsilon_{zz} - \alpha_{zz}T \\ \gamma_{yz} \\ \gamma_{xz} \\ \gamma_{xy} \end{bmatrix} - \begin{bmatrix} 0 & 0 & e_{31} \\ 0 & 0 & e_{32} \\ 0 & 0 & e_{33} \\ 0 & e_{24} & 0 \\ e_{15} & 0 & 0 \\ 0 & 0 & 0 \end{bmatrix} \begin{Bmatrix} E_x \\ E_y \\ E_z \end{Bmatrix}, \tag{3.1}
$$

$$
\begin{bmatrix} D_x \\ D_y \\ D_z \end{bmatrix} = \begin{bmatrix} 0 & 0 & 0 & 0 & e_{15} & 0 \\ 0 & 0 & 0 & e_{24} & 0 & 0 \\ e_{31} & e_{32} & e_{33} & 0 & 0 & 0 \end{bmatrix} \begin{Bmatrix} \varepsilon_{xx} - \alpha_{xx}T \\ \varepsilon_{yy} - \alpha_{yy}T \\ \varepsilon_{zz} - \alpha_{zz}T \\ \gamma_{yz} \\ \gamma_{xz} \\ \gamma_{xy} \end{Bmatrix} + \begin{bmatrix} \in_{11} & 0 & 0 \\ 0 & \in_{22} & 0 \\ 0 & 0 & \in_{33} \end{bmatrix} \begin{Bmatrix} E_x \\ E_y \\ E_z \end{Bmatrix}, \tag{3.2}
$$

where C_{ij}, e_{ij}, and $\in_{ii}$ denote the elastic, piezoelectric, and dielectric constants, respectively; α_{ii} and T are thermal expansion constant and temperature, respectively. By expressing the electric field in terms of electric potential, as outlined by Ghorbanpour Arani et al. [2], we have:

$$
E_k = -\nabla\Phi. \tag{3.3}
$$

Here, the electric potential is assumed as follows:

$$\Phi(x,z,t) = -\cos\left(\frac{\pi z}{h}\right)\phi(x,t) + \frac{2V_0 z}{h}, \tag{3.4}$$

where V_0 represents the external voltage. Consequently, the electric field in two directions can be obtained as follows:

$$E_x = -\frac{\partial \Phi}{\partial x} = \cos\left(\frac{\pi z}{h}\right)\frac{\partial \phi}{\partial x}, \tag{3.5}$$

$$E_z = -\frac{\partial \Phi}{\partial z} = -\frac{\pi}{h}\sin\left(\frac{\pi z}{h}\right)\phi - \frac{2V_0}{h}. \tag{3.6}$$

According to sinusoidal theory, the electromechanical coupling relationship can be concisely written as follows:

$$\sigma_{xx} = C_{11}\left(\varepsilon_{xx} - \alpha_{xx}T\right) + e_{31}\left(\frac{\pi}{h}\sin\left(\frac{\pi z}{h}\right)\phi + \frac{2V_0}{h}\right), \tag{3.7}$$

$$\sigma_{xz} = C_{55}\varepsilon_{xz} - e_{15}\left(\cos\left(\frac{\pi z}{h}\right)\frac{\partial \phi}{\partial x}\right), \tag{3.8}$$

$$D_x = e_{15}\varepsilon_{xz} + \in_{11}\left(\cos\left(\frac{\pi z}{h}\right)\frac{\partial \phi}{\partial x}\right), \tag{3.9}$$

$$D_z = e_{31}\left(\varepsilon_{xx} - \alpha_{xx}T\right) - \in_{33}\left(\frac{\pi}{h}\sin\left(\frac{\pi z}{h}\right)\phi + \frac{2V_0}{h}\right). \tag{3.10}$$

It is important to note that by employing the microelectromechanical model, both the mechanical and electrical properties of the structure can be derived, as demonstrated by Tang and Tong [16]:

$$C_{11} = \frac{C_{11}^r C_{11}^m}{\rho C_{11}^m + (1-\rho)C_{11}^r}, \tag{3.11}$$

$$C_{55} = \frac{1}{\dfrac{\rho \in_{11}^r}{C_{55}^r \in_{11}^r} + \dfrac{(1-\rho)\in_{11}^m}{C_{55}^m \in_{11}^m}}, \tag{3.12}$$

$$e_{11} = Q_{11}\left[\frac{\rho e_{11}^r}{C_{11}^r} + \frac{(1-\rho)e_{11}^m}{C_{11}^m}\right], \tag{3.13}$$

$$e_{15} = \frac{B}{B^2 + AC}, \tag{3.14}$$

$$\in_{11} = \frac{C}{B^2 + AC}, \tag{3.15}$$

$$\in_{33} = \rho \in_{33}^{p} + (1-\rho) \in_{33}^{m} - \frac{e_{31}^2}{C_{11}} + \frac{\rho(e_{31}^{p})^2}{C_{11}^{p}} + \frac{(1-\rho)(e_{31}^{m})^2}{C_{11}^{m}}, \tag{3.16}$$

where

$$A = \frac{\rho C_{55}^{p}}{\left(e_{15}^{p}\right)^2 + C_{55}^{p} \in_{11}^{p}} + \frac{(1-\rho)C_{55}^{m}}{\left(e_{15}^{m}\right)^2 + C_{55}^{m} \in_{11}^{m}}, \tag{3.17}$$

$$B = \frac{\rho e_{15}^{p}}{\left(e_{15}^{p}\right)^2 + C_{55}^{p} \in_{11}^{p}} + \frac{(1-\rho)e_{15}^{m}}{\left(e_{15}^{m}\right)^2 + C_{55}^{m} \in_{11}^{m}}, \tag{3.18}$$

$$C = \frac{\rho \in_{11}^{p}}{\left(e_{15}^{p}\right)^2 + C_{55}^{p} \in_{11}^{p}} + \frac{(1-\rho)\in_{11}^{m}}{\left(e_{15}^{m}\right)^2 + C_{55}^{m} \in_{11}^{m}}. \tag{3.19}$$

Superscripts r and m denote the reinforced and matrix components of the composite, respectively. Furthermore, the symbol ρ indicates the volume percentage of ZnO nanoparticles incorporated into the concrete mixture.

The potential energy of the beam may be expressed as:

$$U = \frac{1}{2} \int_{V} \left(\sigma_{xx}\varepsilon_{xx} + \sigma_{xz}\varepsilon_{xz} - D_x E_x - D_z E_z \right) dV. \tag{3.20}$$

Submitting Eqs. (1.20) and (1.21) as well as Eqs. (3.5) and (3.6) into Eq. (3.20), we have:

$$U = \int_{x} \left[N_x \frac{\partial u}{\partial x} + \frac{N_x}{2}\left(\frac{\partial w}{\partial x}\right)^2 - M_x \frac{\partial^2 w}{\partial x^2} + P_x \frac{\partial \psi}{\partial x} + Q_x \psi \right] dx$$
$$+ \int_{V} \left[-D_x \left(\cos\left(\frac{\pi z}{h}\right)\frac{\partial \phi}{\partial x} \right) - D_z \left(-\frac{\pi}{h}\sin\left(\frac{\pi z}{h}\right)\phi - \frac{2V_0}{h} \right) \right] dV, \tag{3.21}$$

where the resultant force (N_x, Q_x) and bending moment M_x, are:

$$N_x = \int_{A} \sigma_{xx} \, dA, \tag{3.22}$$

$$M_x = \int_{A} \sigma_{xx} z \, dA, \tag{3.23}$$

$$Q_x = \int_{A} \sigma_{xz} \, dA. \tag{3.24}$$

The force induced by the elastic foundation with spring (k_w) and shear (k_g) constants may be denoted as:

$$F_{Elastic\,Medium} = -k_w W + k_g \nabla^2 W. \tag{3.25}$$

Based on the Hamilton's principle, we have:

$$\int_0^t (\delta U - \delta W)dt = 0, \tag{3.26}$$

where

$$\delta u : \frac{\partial N_x}{\partial x} = 0, \tag{3.27}$$

$$\delta w : \frac{\partial^2 M_x}{\partial x^2} - \frac{\partial}{\partial x}\left(N_x^M \frac{\partial w}{\partial x}\right) - k_w w + k_g \nabla^2 w = 0, \tag{3.28}$$

$$\delta \psi : \frac{\partial P_x}{\partial x} - Q_x = 0, \tag{3.29}$$

$$\delta \phi : \int_{-h/2}^{h/2} \left[\frac{D_z \pi}{h}\sin\left(\frac{\pi z}{h}\right) + \cos\left(\frac{\pi z}{h}\right)\frac{\partial D_x}{\partial x}\right] dz = 0, \tag{3.30}$$

where N_x^M is the mechanical axial load. By substituting Eqs. (3.7) to (3.10) into Eqs. (3.22) to (3.24), we have:

$$N_x = hC_{11}\frac{\partial u}{\partial x} + \frac{hC_{11}}{2}\left(\frac{\partial w}{\partial x}\right)^2 + 2e_{31}V_0, \tag{3.31}$$

$$M_x = -C_{11}I\frac{\partial^2 w}{\partial x^2} + \frac{24C_{11}I}{\pi^3}\frac{\partial \psi}{\partial x}, \tag{3.32}$$

$$P_x = -\frac{24C_{11}I}{\pi^3}\frac{\partial^2 w}{\partial x^2} + \frac{6C_{11}I}{\pi^2}\frac{\partial \psi}{\partial x} + \frac{e_{31}h}{2}\varphi, \tag{3.33}$$

$$Q_x = \frac{C_{55}A}{2}\psi - \frac{e_{15}h}{2}\frac{\partial \phi}{\partial x}, \tag{3.34}$$

where

$$(A, I) = \int_A \left(1, z^2\right) dA. \tag{3.35}$$

By substituting Eqs. (3.31) to (3.34) into (3.27) to (3.30), we have:

$$\delta u : \frac{\partial^2 u}{\partial x^2} + \frac{\partial w}{\partial x}\frac{\partial^2 w}{\partial x^2} = 0, \tag{3.36}$$

$$\delta w : -Q_{11}I\frac{\partial^4 w}{\partial x^4} + \frac{24Q_{11}I}{\pi^3}\frac{\partial^3 \psi}{\partial x^3} - \left(2e_{31}V_0 + N_x^M\right)\frac{\partial^2 w}{\partial x^2} - k_w w + k_g \nabla^2 w = 0, \tag{3.37}$$

$$\delta \psi : -\frac{24Q_{11}I}{\pi^3}\frac{\partial^3 w}{\partial x^3} + \frac{6Q_{11}I}{\pi^2}\frac{\partial^2 \psi}{\partial x^2} + \frac{e_{31}h}{2}\frac{\partial \phi}{\partial x} - \frac{Q_{55}A}{2}\psi + \frac{e_{15}h}{2}\frac{\partial \phi}{\partial x} = 0, \tag{3.38}$$

$$\delta \phi : -\frac{2h}{\pi}\frac{\partial^2 w}{\partial x^2} + \frac{h}{2}\frac{\partial \psi}{\partial x} - \frac{\pi^2 \in_{33}}{2h}\phi + \frac{h}{2}\frac{\partial \psi}{\partial x} + \frac{h\in_{11}}{2}\frac{\partial^2 \phi}{\partial x^2} = 0. \tag{3.39}$$

Boundary condition equations are:

- **Clamped–Clamped (CC)**

$$w = u = \phi = \psi = \frac{\partial w}{\partial x} = 0, \qquad @ \quad x = 0$$
$$w = u = \phi = \psi = \frac{\partial w}{\partial x} = 0. \qquad @ \quad x = L \tag{3.40}$$

- **Clamped–Simple (CS)**

$$w = u = \phi = \psi = \frac{\partial w}{\partial x} = 0, \qquad @ \quad x = 0$$
$$w = u = \phi = \frac{\partial \psi}{\partial x} = \frac{\partial^2 w}{\partial x^2} = 0. \qquad @ \quad x = L \tag{3.41}$$

- **Simple–Simple (SS)**

$$w = u = \phi = \frac{\partial \psi}{\partial x} = \frac{\partial^2 w}{\partial x^2} = 0, \qquad @ \quad x = 0$$
$$w = u = \phi = \frac{\partial \psi}{\partial x} = \frac{\partial^2 w}{\partial x^2} = 0. \qquad @ \quad x = L \tag{3.42}$$

Due to the presence of weight coefficients, the boundary condition equations become coupled, leading to a complex interdependency. Consequently, it is necessary to separate the boundary conditions from the governing equations. Based on the DQM, presented in Chapter 2, the resulting system of equations, including both the governing equations and the boundary conditions, can be expressed in matrix form as follows:

$$\left(\left[\underbrace{K_I + K_{NI}}_{K}\right] + P\left[K_g\right]\right) \begin{Bmatrix} \{d_b\} \\ \{d_d\} \end{Bmatrix} = 0, \tag{3.43}$$

where P is the buckling load; $[K_L]$, $[K_{NL}]$, and $[K_g]$, respectively, are the linear stiffness matrix, nonlinear stiffness matrix, and geometric matrix. Hence, based on the eigenvalue problem, the buckling load may be determined.

3.3 NUMERICAL RESULTS

In this section, we consider a beam with an elastic modulus of $E_m = 18$ GPa, which is reinforced with ZnO nanoparticles possessing an elastic modulus of $E_r = 130$ GPa.

3.3.1 ACCURACY OF THE DQM

Figure 3.2 illustrates the relationship between the structural buckling load of the column and the number of grid points. It is evident that as the number of grid points increases, the buckling load gradually decreases until it converges at a certain point. Therefore, for this chapter, calculations are conducted using 15 grid points.

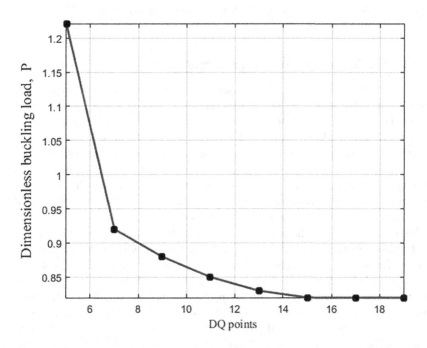

FIGURE 3.2 Accuracy of the DQM for buckling load.

TABLE 3.1

Validation of the Present Work with Other Published Works

L/h	SSDT, Thai and Vo [17]	Present work
5	8.9533	8.9534
10	9.6232	9.6233
20	9.8068	9.8069
100	9.8671	9.8675

3.3.2 VALIDATION

No researcher has previously investigated the buckling behavior of concrete columns reinforced with ZnO nanoparticles. Therefore, to validate our findings and isolate the effects of zinc oxide nanoparticles $(\rho = 0)$, foundations $(k_w = k_g = 0)$, and piezo-electric properties, we conducted a buckling analysis of a beam utilizing the sinusoidal model. Adopting material and geometric parameters akin to those used by Thai and Vo [17], we presented the buckling load for various aspect ratios of the structure in Table 3.1. It is evident from the comparison that our results align well with those reported by Thai and Vo [17], indicating the accuracy of our findings. It should be noted that any slight discrepancies between our results and those of Thai and Vo [17] are primarily attributable to differences in the underlying theories employed. Specifically, while we utilized the SSDT model in this chapter, Thai and Vo [17] applied the Timoshenko beam theory.

3.3.3 THE EFFECT OF DIFFERENT PARAMETERS

Figure 3.3 depicts the impact of the volume percentage of ZnO nanoparticles on the dimensionless buckling load of the structure as a function of the number of longitudinal modes. The point at which the buckling load is minimized is referred to as the critical buckling load, which, in this case, occurs at the third longitudinal mode. Additionally, it is observed that increasing the volume fraction of nanoparticles leads to the augmentation of the buckling load of the column. This phenomenon can be attributed to the increased presence of nanoparticles within the columns, resulting in a reduction in surface area. Furthermore, it is inferred that the effects of ZnO nanoparticles become more pronounced at higher modes.

Figure 3.4 illustrates the impact of the foundation on the dimensionless buckling load across various longitudinal modes. Three types of foundations are examined: without foundation, foundation modeled with vertical springs (Winkler), and foundation modeled with vertical springs and shear layer (Pasternak). It is evident that incorporating a foundation leads to an increase in the buckling load of the column. Furthermore, the buckling load of the column with a Pasternak foundation exceeds that of the column with a Winkler foundation. This discrepancy can be attributed to the additional consideration of shear force in the Pasternak model, alongside the flexibility factor.

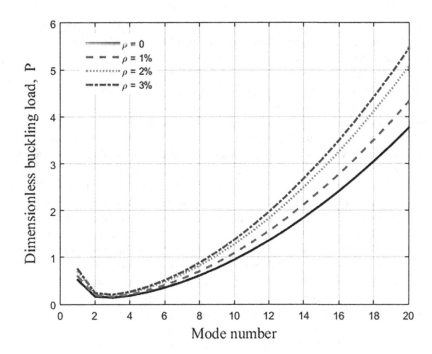

FIGURE 3.3 The impact of ZnO nanoparticle volume percent on the buckling load.

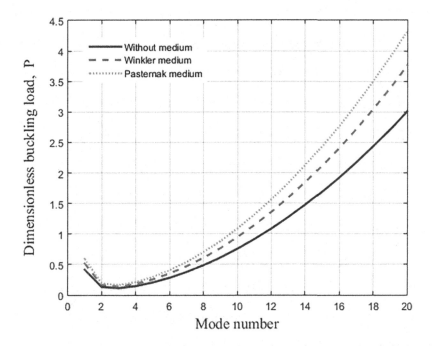

FIGURE 3.4 The impact of the elastic medium on the buckling load.

Figure 3.5 illustrates the impact of the dimensionless external applied voltage $\left(V^* = V_0 / (h\sqrt{E_r / \in_{11}}\right)$ on the dimensionless buckling load of columns across various longitudinal modes. It is observed that a negative external voltage leads to a compressive force and an increase in the buckling load of the system. Conversely, this phenomenon is reversed for a positive external voltage. Nevertheless, it can be inferred that the application of external voltage serves as an effective controlling parameter for the buckling behavior of concrete columns.

Figure 3.6 demonstrates the impact of different boundary conditions on the buckling load across various longitudinal modes. The choice of boundary conditions plays a pivotal role in determining the structural response to external forces. Notably, in columns with clamped boundary conditions at both ends, the buckling load reaches its maximum value compared to other boundary condition configurations. This is attributed to the clamped–clamped (CC) boundary condition, which imposes the highest level of structural stiffness among the considered cases. Additionally, it is observed that the buckling load of a column with clamped–simple boundary conditions surpasses that of a column with simple–simple boundary conditions. This suggests that the introduction of fixed ends leads to a higher resistance against buckling compared to the scenario where both ends are simply supported. Overall, these findings underscore the importance of selecting appropriate boundary conditions to optimize the buckling behavior of the structure.

Figure 3.7 illustrates the influence of temperature on the dimensionless buckling load of the structure across various longitudinal modes. The point at which the

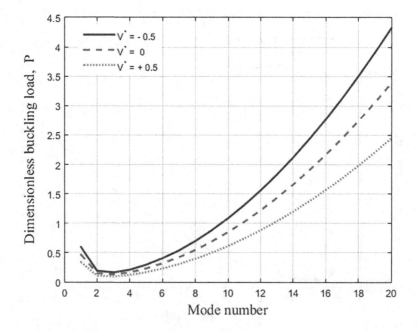

FIGURE 3.5 The impact of external applied voltage on the buckling load.

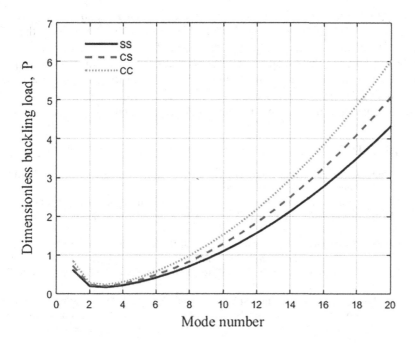

FIGURE 3.6 The impact of boundary conditions on the buckling load.

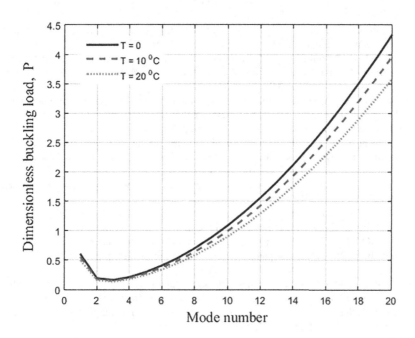

FIGURE 3.7 The impact of temperature on the buckling load.

buckling load is minimized is referred to as the critical buckling load, which, in this case, occurs at the third longitudinal mode. It is observed that increasing the temperature leads to a decrease in the buckling load of the column. This phenomenon can be attributed to the reduction in stiffness that occurs with increasing temperature. As the temperature rises, the material properties of the column are affected, resulting in a decrease in its resistance to buckling. Consequently, understanding the thermal behavior of the structure is crucial for mitigating buckling risks.

REFERENCES

1. Formica G, Lacarbonara W, Alessi R. Vibrations of carbon nanotube reinforced composites. *J. Sound Vib.* 2010;329:1875–1889.
2. Ghorbanpour Arani A, Kolahchi R, Vossough H. Buckling analysis and smart control of SLGS using elastically coupled PVDF nanoplate based on the nonlocal Mindlin plate theory. *Physica B.* 2012;407:4458–4466.
3. Ghorbanpour Arani A, Fereidoon A, Kolahchi R. Nonlinear surface and nonlocal piezoelasticity theories for vibration of embedded single-layer boron nitride sheet using harmonic differential quadrature and differential cubature methods. *J. Intell. Mat. Syst. Struct.* 2014;17:1–12.
4. Ghorbanpour Arani A, Haghparast E, Khoddami Maraghi Z, Amir S. Static stress analysis of carbon nano-tube reinforced composite (CNTRC) cylinder under non-axisymmetric thermo-mechanical loads and uniform electro-magnetic fields. *Compos. Part B: Eng.* 2015;68:136–145.
5. Henkhaus K, Pujol S, Ramirez J. Axial failure of reinforced concrete columns damaged by shear reversals. *J. Struct. Eng.* 2013;73:1172–1180.
6. Jafarian Arani A, Kolahchi R. Buckling analysis of embedded concrete columns armed with carbon nanotubes. *Comput. Concr.* 2016;17:567–578.
7. Kadoli R, Ganesan N. Free vibration and buckling analysis of composite cylindrical shells conveying hot fluid. *Compos. Struct.* 2003;60:19–32.
8. Kolahchi R, Rabani Bidgoli M, Beygipoor Gh, Fakhar MH. A nonlocal nonlinear analysis for buckling in embedded FG-SWCNT-reinforced microplates subjected to magnetic field. *J. Mech. Sci. Tech.* 2013;5:2342–2355.
9. Kolahchi R, Safari M, Esmailpour M. Dynamic stability analysis of temperature-dependent functionally graded CNT-reinforced visco-plates resting on orthotropic elastomeric medium. *Compos. Struct.* 2016a;150:255–265.
10. Kolahchi R, Hosseini H, Esmailpour M. Differential cubature and quadrature-Bolotin methods for dynamic stability of embedded piezoelectric nanoplates based on visco-nonlocal-piezoelasticity theories. *Compos. Struct.* 2016b;157:174–186.
11. Karaca Z, Türkeli E. The slenderness effect on wind response of industrial reinforced concrete chimneys. *Wind Struct.* 2014;18:281–294.
12. Kolahchi R, Moniribidgoli AM. Size-dependent sinusoidal beam model for dynamic instability of single-walled carbon nanotubes. *Appl. Math. Mech.* 2016c;37(2):265–274.
13. Liew KM, Lei ZX, Yu JL, Zhang LW. Postbuckling of carbon nanotube-reinforced functionally graded cylindrical panels under axial compression using a meshless approach. *Comput. Meth. Appl. Mech. Eng.* 2014;268:1–17.
14. Matsuna H. Vibration and buckling of cross-ply laminated composite circular cylindrical shells according to a global higher-order theory. *Int. J. Mech. Sci.* 2007;49:1060–1075.
15. Mirza S, Skrabek B. Reliability of short composite beam column strength interaction. *J. Struct. Eng.* 1991;117(8):2320–2339.

16. Tan P, Tong L. Micro-electromechanics models for piezoelectric-fiber-reinforced composite materials. *Compos. Sci. Tech.* 2001;61:759–769.

17. Thai HT, Vo TP. A nonlocal sinusoidal shear deformation beam theory with application to bending, buckling, and vibration of nanobeams. *Int. J. Eng. Sci.* 2012;54:58–66.

18. Thai HT. A nonlocal beam theory for bending, buckling, and vibration of nanobeams. *Int. J. Eng. Sci.* 2012;52:56–64.

19. Solhjoo S, Vakis AI. Single asperity nanocontacts: Comparison between molecular dynamics simulations and continuum mechanics models. *Comput. Mat. Sci.* 2015;99:209–220.

20. Seo YS, Jeong WB, Yoo WS, Jeong HK. Frequency response analysis of cylindrical shells conveying fluid using finite element method. *J. Mech. Sci. Tech.* 2015;19:625–633.

21. Wuite J, Adali S. Deflection and stress behaviour of nanocomposite reinforced beams using a multiscale analysis. *Compos. Struct.* 2005;71:388–396.

22. Zamanian M, Kolahchi R, Rabani Bidgoli M. Agglomeration effects on the buckling behaviour of embedded concrete columns reinforced with SiO_2 nanoparticles. *Wind Struct.* 2017;24:43–57.

4 Vibration of Nanocomposite Beams

4.1 INTRODUCTION

Reinforced concrete (RC) is a composite material with a reinforcement, which can be steel bars, plates, fibers, or nanoparticles. Recently, the usage of different types of nanoparticles in concrete structures has been an intense interest among researchers, since the nanoparticles can improve the quality and material properties of concrete. With respect to the fact that the nanoparticles can be agglomerated in the concrete, however, in this chapter, a mathematical model is introduced for a concrete beam reinforced with nanoparticles to estimate the vibration behavior of the mentioned structures considering agglomeration effects.

With respect to the developed works in the field of RC structures, Kim and Aboutaha [1] conducted a study utilizing a three-dimensional nonlinear finite element model to analyze the behavior of reinforced concrete beams strengthened with carbon fiber-reinforced polymer (CFRP) composites, focusing on enhancing the beams' flexural capacity and ductility. Nazari and Riahi [2] presented evaluations of strength and water absorption coefficients in high-performance self-compacting concrete containing varying amounts of TiO_2 nanoparticles. Khalaj and Nazari [3] explored the split tensile strength of self-compacting concrete mixed with SiO_2 nanoparticles and various quantities of randomly oriented steel fibers. Khoshakhlagh et al. [4] investigated the compressive, flexural, and split tensile strengths, along with the water absorption coefficient, of high-performance self-compacting concrete with different amounts of Fe_2O_3 nanoparticles. Jalala et al. [5] examined the enhancement in strength and durability-related characteristics, as well as the rheological, thermal, and microstructural properties of high-strength self-compacting concrete (HSSCC) containing nano TiO_2 and industrial waste fly ash (FA). Ibraheem et al. [6] identified the addition of steel fibers to the concrete mixture as a novel mass reinforcement technique that improves torsional, flexural, and shear behaviors of structural members. El-Helou and Aboutaha [7] analyzed the nominal moment-axial load interaction diagrams, moment-curvature relationships, and ductility of rectangular hybrid beam–column concrete sections using the modified Hognestad concrete model. Güneyisi et al. [8] investigated the physico-mechanical properties of self-compacting lightweight aggregate concrete (SCLC) incorporating artificial lightweight aggregate (LWA) made from fly ash (FA) through a cold-bonding process. Ibraheem [9] conducted an experimental investigation into the behavior and cracking of steel fiber reinforced concrete spandrel L-shaped beams under combined torsion, bending, and shear. Le et al. [10] presented experimental results for three large-scale concrete composite beams featuring a new puzzle-shaped crestbond. Saribiyik and Caglar [11] produced RC

DOI: 10.1201/9781003510710-4

specimens to evaluate RC beams with inadequate shear and tensile reinforcement made using low-strength concrete. Ding et al. [12] investigated the flexural stiffness of simply supported steel-concrete composite I-beams under positive bending moments through combined experimental, numerical, and standard methods. Hind et al. [13] performed numerical analysis on a series of concrete beams reinforced with short fibers, utilizing a database of experimental results from existing literature.

The mathematical modeling of concrete structures is a novel topic that has recently become an intense point of interest among researchers. Jafarian Arani et al. [14] and Zamanian et al. [15] studied the nonlinear buckling behavior of straight concrete columns reinforced with single-walled carbon nanotubes (SWCNTs) and SiO_2 nanoparticles, all supported on a foundation. Additionally, Heidarzadeh et al. [16] focused on the nonlinear buckling of straight concrete columns reinforced solely with SWCNTs, also resting on a foundation. Stress analysis of concrete pipes reinforced with AL_2O_3 nanoparticles was presented by Heidarzadeh et al. (2016) considering agglomeration effects.

To the best of our knowledge, no theoretical report has been found in the literature on the vibration analysis of concrete beams reinforced with nanoparticles. Motivated by these considerations, we aim to present a mathematical model for the vibration analysis of embedded concrete columns reinforced with SiO_2 nanoparticles considering agglomeration effects based on the Mori–Tanaka approach. Based on Timoshenko beam model, the motion equations are derived using the energy method and Hamilton's principal. Using the DQM, the frequency of the structure is calculated, and the effects of different parameters such as volume percent of SiO_2 nanoparticles, SiO_2 agglomeration, geometrical parameters, elastic foundation, and boundary conditions on the frequency of concrete beam are shown.

4.2 MOTION EQUATIONS

Figure 4.1 shows an embedded concrete beam reinforced with agglomerated SiO_2 nanoparticles. The surrounding foundation is described by the Pasternak model containing the spring and shear constants.

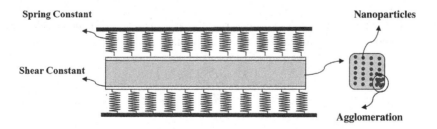

FIGURE 4.1 Schematic of an embedded concrete column reinforced with agglomerated SiO_2 nanoparticles.

By applying the Timoshenko beam model, the displacements fields and strain relations based on section 1.3.2 of Chapter 1 are assumed. Based on Hook's low, the isotropic stress–strain relations can be written as:

$$\sigma_{xx} = \frac{E}{1-\nu^2}\left[\frac{\partial u}{\partial x} + z\frac{\partial \psi}{\partial x} + \frac{1}{2}\left(\frac{\partial w}{\partial x}\right)^2\right], \tag{4.1}$$

$$\sigma_{xz} = \frac{E}{2(1+\nu)}\left[\frac{\partial W}{\partial x} + \psi\right], \tag{4.2}$$

where E and ν are Young's modulus and Poison's ratio of the SiO_2-reinforced concrete beam, which can be calculated by the Mori–Tanaka model as [17]:

$$E = \frac{9KG}{3K+G}, \tag{4.3}$$

$$\upsilon = \frac{3K-2G}{6K+2G}, \tag{4.4}$$

where the effective bulk modulus (K) and effective shear modulus (G) may be expressed as:

$$K = K_{out}\left[1 + \frac{\xi\left(\frac{K_{in}}{K_{out}}-1\right)}{1+\alpha(1-\xi)\left(\frac{K_{in}}{K_{out}}-1\right)}\right], \tag{4.5}$$

$$G = G_{out}\left[1 + \frac{\xi\left(\frac{G_{in}}{G_{out}}-1\right)}{1+\beta(1-\xi)\left(\frac{G_{in}}{G_{out}}-1\right)}\right], \tag{4.6}$$

where the two parameters ξ and ζ describe the agglomeration of nanoparticles, and C_r is related to the SiO_2 volume fraction. In addition, $\chi_r, \beta_r, \delta_r, \eta_r$ may be calculated as:

$$\chi_r = \frac{3(K_m+G_m)+k_r-l_r}{3(k_r+G_m)}, \tag{4.7}$$

$$\beta_r = \frac{1}{5}\left[\frac{4G_m+2k_r+l_r}{3(k_r+G_m)} + \frac{4G_m}{(p_r+G_m)} + \frac{2[G_m(3K_m+G_m)+G_m(3K_m+7G_m)]}{G_m(3K_m+G_m)+m_r(3K_m+7G_m)}\right], \tag{4.8}$$

$$\delta_r = \frac{1}{3}\left[n_r + 2l_r + \frac{(2k_r - l_r)(3K_m + 2G_m - l_r)}{k_r + G_m} \right],$$ (4.9)

$$\eta_r = \frac{1}{5}\left| \begin{array}{l} \frac{2}{3}(n_r - l_r) + \frac{4G_m P_r}{(P_r + G_m)} + \frac{8G_m m_r (3K_m + 4G_m)}{3K_m (m_r + G_m) + G_m (7m_r + G_m)} \\ + \frac{2(k_r - l_r)(2G_m + l_r)}{3(k_r + G_m)} \end{array} \right|,$$ (4.10)

where k_r, l_r, n_r, p_r, and m_r are the Hill's elastic moduli for the nanoparticles; K_m and G_m are the bulk and shear moduli of the matrix, which can be written as:

$$K_m = \frac{E_m}{3(1 - 2\upsilon_m)},$$ (4.11)

$$G_m = \frac{E_m}{2(1 + \upsilon_m)},$$ (4.12)

where E_m and υ_m are Young's modulus and the Poisson's ratio of the concrete beam, respectively. Furthermore, β, α can be obtained from:

$$\alpha = \frac{(1 + \upsilon_{out})}{3(1 - \upsilon_{out})},$$ (4.13)

$$\beta = \frac{2(4 - 5\upsilon_{out})}{15(1 - \upsilon_{out})},$$ (4.14)

$$\upsilon_{out} = \frac{3K_{out} - 2G_{out}}{6K_{out} + 2G_{out}}.$$ (4.15)

The strain energy of the nanocomposite concrete beam can be expressed as

$$U = \frac{1}{2}\int_0^L \int_A (\sigma_{xx}\varepsilon_{xx} + \sigma_{xz}\gamma_{xz}) \, dV.$$ (4.16)

Substituting Eqs. (4.1) and (4.2) into (4.16) yields:

$$U = \frac{1}{2}\int_0^L \int_A \left[\sigma_{xx}\left(\frac{\partial u}{\partial x} + z\frac{\partial \psi}{\partial x} + \frac{1}{2}\left(\frac{\partial w}{\partial x}\right)^2 \right) + \sigma_{xz}\left(\frac{\partial w}{\partial x} + \psi \right) \right] dV.$$ (4.17)

The kinetic energy of the structure can be written as:

$$K = \frac{\rho}{2} \int_0^L \int_A \left[(\dot{U}_1)^2 + (\dot{U}_2)^2 + (\dot{U}_3)^2 \right] dV, \tag{4.18}$$

where ρ is the density of the nanocomposite concrete beam. Substituting the deflection field into Eq. (4.18) yields:

$$K = \frac{\rho}{2} \int_0^L \int_A \left[\left(\frac{\partial u}{\partial t} + z \frac{\partial \psi}{\partial t} \right)^2 + \left(\frac{\partial w}{\partial t} \right)^2 \right] dV. \tag{4.19}$$

The external work due to the surrounding foundation can be expressed as [18]:

$$W = \int_0^L \left(-k_w w + k_g \nabla^2 w \right) w dx, \tag{4.20}$$

where k_w and k_g are spring and shear constants of foundation, respectively. Hamilton's principle is applied as follows:

$$\int_0^t \left(\delta U - \delta K - \delta W \right) dt = 0. \tag{4.21}$$

The motion equations of the structure can be derived as follows:

$$\frac{Eh}{1-\nu^2} \left(\frac{\partial^2 u}{\partial x^2} + \frac{\partial^2 w}{\partial x^2} \frac{\partial w}{\partial x} \right) = \rho h \frac{\partial^2 u}{\partial t^2}, \tag{4.22}$$

$$\frac{K_s Eh}{2(1+\nu)} \left[\frac{\partial^2 w}{\partial x^2} + \frac{\partial \psi}{\partial x} \right] - k_w w + k_g \frac{\partial^2 w}{\partial x^2} = \rho h \frac{\partial^2 w}{\partial t^2}, \tag{4.23}$$

$$\frac{Eh^3}{12(1-\nu^2)} \frac{\partial^2 \psi}{\partial x^2} - \frac{K_s Eh}{2(1+\nu)} \left[\frac{\partial w}{\partial x} + \psi \right] = \frac{\rho h^3}{12} \frac{\partial^2 \psi}{\partial t^2}, \tag{4.24}$$

where K_s is the shear correction factor. The associated boundary conditions can be expressed as:

- **Clamped–Clamped Boundary Condition (CC)**

$$w = u = \psi = \frac{\partial w}{\partial x} = 0, \qquad @ \ x = 0$$

$$w = u = \psi = \frac{\partial w}{\partial x} = 0. \qquad @ \ x = L \tag{4.25}$$

- **Clamped–Simple Boundary Condition (CS)**

$$w = u = \psi = \frac{\partial w}{\partial x} = 0, \qquad @ \quad x = 0$$

$$w = u = \frac{\partial \psi}{\partial x} = \frac{\partial^2 w}{\partial x^2} = 0. \qquad @ \quad x = L$$

(4.26)

- **Simple–Simple Boundary Condition (SS)**

$$w = u = \frac{\partial \psi}{\partial x} = \frac{\partial^2 w}{\partial x^2} = 0, \qquad @ \quad x = 0$$

$$w = u = \frac{\partial \psi}{\partial x} = \frac{\partial^2 w}{\partial x^2} = 0. \qquad @ \quad x = L$$

(4.27)

4.3 NUMERICAL RESULTS

Herein, based on the DQM Presented in chapter 2, a concrete column with length of $L = 3$ m, thickness of $h = 30$ cm, Young's modulus of $E_m = 20$ GPa and Poison's ratio of $\nu_m = 0.3$ is considered, which is reinforced with agglomerated SiO_2 nanoparticles with a Young's modulus of $E_r = 75$ GPa and Poison's ratio of $\nu_r = 0.3$.

The effect of the grid point number in the DQM on the frequency of the concrete column is shown in Figure 4.2. As can be seen, the fast rate of convergence of the

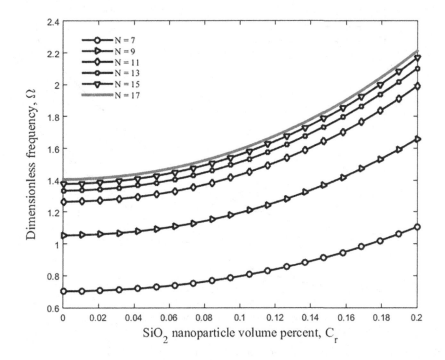

FIGURE 4.2 Accuracy and convergence of the DQM.

method is quite evident, and it is found that 17 DQM grid points can yield accurate results. It can be seen that with increasing the volume percent of SiO$_2$ nanoparticles, the nonlinear frequency increases. This is due to the fact that with an increasing volume percent of SiO$_2$ nanoparticles, the stiffness of the structure increases.

The effects of agglomeration (ξ) on the frequency of the structure versus the SiO$_2$ nanoparticle volume percent are demonstrated in Figure 4.3. It can be seen that considering agglomeration effects leads to lower frequency.

This is due to the fact that considering the agglomeration effect leads to lower stiffness in the structure. However, the agglomeration effect has a major effect on the vibration behavior of the structure. In addition, with increasing C$_r$, the frequency is increased for both models.

Figure 4.4 illustrates the influence of boundary conditions on the frequency along the volume percent of SiO$_2$ nanoparticles. It can be concluded that the frequency is higher for CC boundary conditions with respect to other types of considered boundaries. This is because in the CC boundary condition, the structure becomes stiffer.

The effect of thickness-to-length ratio of the concrete beam on the frequency versus the volume percent of SiO$_2$ nanoparticles is depicted in Figure 4.5. As can be seen, with an increasing thickness-to-length ratio of the concrete beam, the frequency is increased since the stiffness of the structure enhances.

Figure 4.6 presents the influence of the elastic medium on the frequency along the volume percent of SiO$_2$ nanoparticles. Obviously, the foundation has a significant effect on the frequency of the beam, since the frequency of the system in the

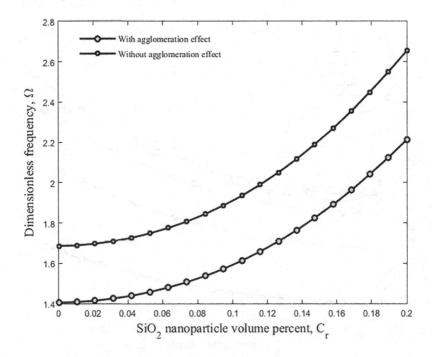

FIGURE 4.3 Effects of agglomeration on the frequency of the structure.

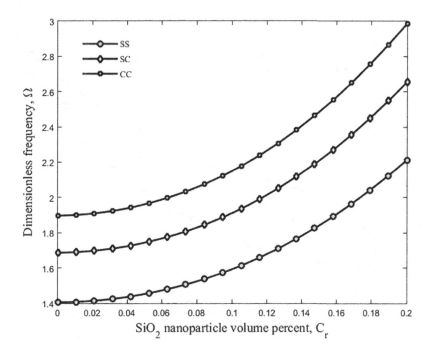

FIGURE 4.4 Effects of boundary condition on the frequency of the structure.

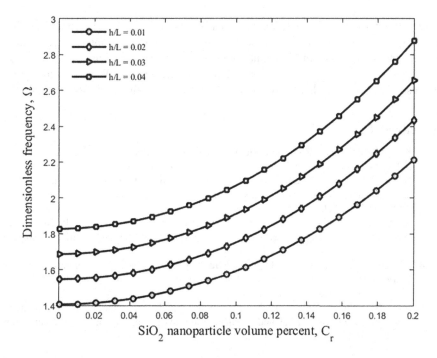

FIGURE 4.5 Effects of thickness-to-length ratio on the frequency of the structure.

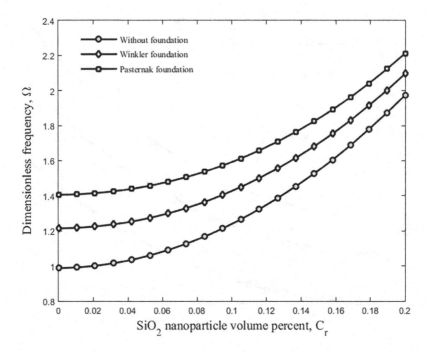

FIGURE 4.6 Effects of elastic medium on the frequency of the structure.

case without a foundation is lower than that of other cases. It can be concluded that the frequency for the Pasternak model (spring and shear constants) is higher than that for the Winkler (spring constant) model. These results are reasonable, since the Pasternak medium considers not only the normal stresses (i.e., Winkler foundation) but also the transverse shear deformation and continuity among the spring elements.

REFERENCES

1. Kim SH, Aboutaha RS. Finite element analysis of carbon fiber-reinforced polymer (CFRP) strengthened reinforced concrete beams. *Comput. Concr.* 2004;1:401–416.
2. Nazari A, Riahi S. The effect of TiO_2 nanoparticles on water permeability and thermal and mechanical properties of high strength self-compacting concrete. *Mat. Sci. Eng. A.* 2010;528:756–763.
3. Khalaj Gh, Nazari A. Modeling split tensile strength of high strength self compacting concrete incorporating randomly oriented steel fibers and SiO_2 nanoparticles. *Compos. Part B: Eng.* 2012;43:1887–1892.
4. Khoshakhlagh A, Nazari A, Khalaj Gh. Effects of Fe2O3 nanoparticles on water permeability and strength assessments of high strength self-compacting *Concr. J. Mat. Sci. Tech.* 2012;28:73–82.
5. Jalala M, Fathi M, Farzad M. Effects of fly ash and TiO_2 nanoparticles on rheological, mechanical, microstructural and thermal properties of high strength self compacting concrete. *Mech. Mat.* 2013;61:11–27.

6. Ibraheem OF, Abu Bakar BH, Johari I. Behavior and crack development of fiber-reinforced concrete spandrel beams under combined loading: An experimental study. *Struct. Eng. Mech.* 2015;54:1–17.

7. El-Helou RG, Aboutaha RS. Analysis of rectangular hybrid steel-GFRP reinforced concrete beam columns. *Comput. Concr.* 2015;16:245–260.

8. Güneyisi E, Gesoglu M, Azez OA, Öz HÖ. Physico-mechanical properties of self-compacting concrete containing treated cold-bonded fly ash lightweight aggregates and SiO_2 nano-particles. *Construct. Build. Mat.* 2015;101:1142–1153.

9. Ibraheem OF, Abu Bakar BH, Johari I. Fiber reinforced concrete L-beams under combined loading. *Comput. Concr.* 2014;14:1–18.

10. Le VPN, Bui DV, Chu THV, Kim IT, Ahn JH, Dao DK. Behavior of steel and concrete composite beams with a newly puzzle shape of crestbond rib shear connector: An experimental study. *Struct. Eng. Mech.* 2016;60:1001–1019.

11. Saribiyik A, Caglar N. Flexural strengthening of RC beams with low-strength concrete using GFRP and CFRP. *Struct. Eng. Mech.* 2016;60:825–845.

12. Ding FX, Liu J, Liu XM, Guo FQ, Jiang LZ. Flexural stiffness of steel-concrete composite beam under positive moment. *Steel Compos. Struct.* 2016;20:1369–1389.

13. Hind MK, Mustafa Ö, Talha E, Abdolbaqi MK. Flexural behavior of concrete beams reinforced with different types of fibers. *Comput. Concr.* 2016;18:999–1018.

14. Jafarian Arani A, Kolahchi R. Buckling analysis of embedded concrete columns armed with carbon nanotubes. *Comput. Concr.* 2016;17:567–578.

15. Zamanian M, Kolahchi R, Rabani Bidgoli M. Agglomeration effects on the buckling behaviour of embedded concrete columns reinforced with SiO_2 nano-particles. *Wind Struct.* 2016;24:43–57.

16. Heidarzadeh A, Kolahchi R, Rabani Bidgoli M. Concrete pipes reinforced with AL_2O_3 nanoparticles considering agglomeration: Magneto-thermo-mechanical stress analysis. *Int. J. Civil. Eng.* 2016. https://doi.org/10.1007/s40999-016-0130-2.

17. Kolahchi R, Safari M, Esmailpour M. Dynamic stability analysis of temperature-dependent functionally graded CNT-reinforced visco-plates resting on orthotropic elastomeric medium. *Compos. Struct.* 2016;150:255–265.

18. Kolahchi R, Rabani Bidgoli M, Beygipoor Gh, Fakhar MH. A nonlocal nonlinear analysis for buckling in embedded FG-SWCNT-reinforced microplates subjected to magnetic field. *J. Mech. Sci. Tech.* 2015;29:3669–3677.

5 Dynamic Buckling of Nanocomposite Beams

5.1 INTRODUCTION

Sandwich structures can be used in different industries such as aerospace, aircraft, automobile, etc. due to high strength and low weight with respect to traditional materials. One of the important ways to control sandwich structures is using piezoelectric materials, since in these materials, the structure subjected to mechanical forces can produce an electric field and vice versa [1, 2].

The dynamic analysis of sandwich structures has been reported by researchers. An investigation on the nonlinear dynamic response and vibration of the imperfect laminated three-phase polymer nanocomposite panel resting on elastic foundations was presented by Duc et al. [3]. Van Thu and Duc [4] presented an analytical approach to investigate the non-linear dynamic response and vibration of an imperfect three-phase laminated nanocomposite cylindrical panel resting on elastic foundations in thermal environments. Shokravi and Jalili [5] studied nonlocal temperature-dependent dynamic buckling analysis of embedded sandwich micro plates reinforced by functionally graded carbon nanotubes. Shokravi [6] presented temperature-dependent buckling analysis of sandwich nanocomposite plates resting on elastic medium subjected to magnetic field. Buckling analysis of embedded laminated plates with nanocomposite layers was studied by Shokravi [7]. Duc et al. [8–11] studied thermal and mechanical stability of a functionally graded composite truncated conical shell, plates and double curved shallow shells reinforced by carbon nanotube fibers. Based on Reddy's third-order shear deformation plate theory, the nonlinear dynamic response and vibration of imperfect functionally graded carbon nanotube-reinforced composite plates was analyzed by Thanh et al. [12]. Duc et al. [11] presented the first analytical approach to investigate the nonlinear dynamic response and vibration of imperfect rectangular nanocompsite multilayer organic solar cell subjected to mechanical loads using the classical plate theory. Katariya et al. [13] reported the thermal buckling strength of the sandwich shell panel structure and subsequent improvement of the same by embedding shape memory alloy (SMA) fibre via a general higher-order mathematical model in conjunction with finite element method. The axisymmetric buckling delamination of the Piezoelectric/Metal/Piezoelectric (PZT/Metal/PZT) sandwich circular plate with interface penny-shaped cracks was investigated by Cafarova et al. [14]. To control the stochastic vibration of a vibration-sensitive instrument supported on a beam, the beam was designed by Ying et al. (2017) as a sandwich structure with magneto-rheological visco-elastomer (MRVE) core. Aerothermoelastic flutter and thermal buckling characteristics of sandwich panels with the pyramidal lattice core resting on elastic foundations in supersonic airflow were studied by Chai et al. [15]. A critical review of literature on bending,

DOI: 10.1201/9781003510710-5

buckling and free vibration analysis of shear deformable isotropic, laminated composite and sandwich beams based on equivalent single layer theories, layerwise theories, zig-zag theories and exact elasticity solution was presented by Sayyad and Ghugal [16]. Song and Li [17] investigated the flutter and buckling properties of sandwich panels with triangular lattice core in supersonic airflow, and the active flutter and buckling control are also carried out, which can provide theoretical basis for the use of sandwich structures in the design of aircrafts. Van Do and Lee [18] reported thermal buckling analyses of functionally graded material (FGM) sandwich plates using an improved mesh-free radial point interpolation method (RPIM). Latifi et al. [19] studied nonlinear dynamic instability analysis of three-layered composite beams with viscoelastic core subjected to combined lateral and axial loadings. Li et al. [20] focused on the post-buckling and free vibration of the sandwich beam theoretically in thermal environments with simply supported and clamped boundary conditions. Shamshuddin Sayyad and Ghugal [21] investigated the bending, buckling, and vibration responses of shear deformable laminated composite and sandwich beams using trigonometric shear and normal deformation theory. Analytical closed-form solutions for thermos-mechanical stability and explicit expressions for free- and forced-vibration of thin functionally graded sandwich shells with double curvature resting on elastic bases were investigated by Trinh and Kim [22].

In this chapter, the dynamic buckling of a sandwich beam with piezoelectric layers subjected to an electric field is studied. The structure is modeled by the hyperbolic shear deformation beam theory (HSDBT), and the motion equations are derived by the energy method. The elastic foundation is simulated by the Pasternak model with spring and shear elements. The DIR of the structure is calculated by the differential quadrature method (DQM) in conjunction with the Bolotin method. The effects of applied voltage, geometrical parameters of the structure, and boundary conditions on the DIR of the structure are shown.

5.2 MOTION EQUATIONS

In Figure 5.1, a sandwich beam subjected to electric field is shown. The length of the structure is L, and the thickness of core, top, and bottom layers are shown by h_c, h_t, and h_b, respectively. The structure is rested on a Pasternak foundation with spring and shear elements.

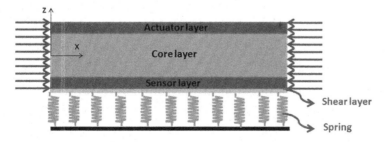

FIGURE 5.1 Schematic of sandwich beam rested on a Pasternak foundation.

Based on hyperbolic shear deformation theory presented in Chapter 2 and Eqs. (1.22) to (1.26), the structure will be modeled. The stress–strain relations of the core layer are:

$$\sigma_{xx}^c = C_{11}\varepsilon_{xx}, \tag{5.1}$$

$$\tau_{xz}^c = C_{44}\gamma_{xz}, \tag{5.2}$$

where C_{11} and C_{44} are the elastic constants of the core. Also, based on piezoelasticity theory, the constitutive relation of the piezoelectric facesheets can be expressed as [23]:

$$\sigma_{xx}^p = Q_{11}\varepsilon_{xx} - e_{31}E_z, \tag{5.3}$$

$$\tau_{xz}^p = Q_{44}\gamma_{xz} - e_{15}E_x, \tag{5.4}$$

$$D_x = e_{15}\varepsilon_{xz} + \in_{11} E_x, \tag{5.5}$$

$$D_z = e_{31}\gamma_{xx} + \in_{33} E_z, \tag{5.6}$$

in which Q_{11} and Q_{44} are the elastic constants; e_{31} and e_{15} are the piezoelectric constants; $\in_{11}$ and $\in_{33}$ are the dielectric constants; D_x and D_z are the electric displacements in the x and z directions, respectively; and E_x and E_z are the electric fields in the x and z directions, respectively which can be defined as [24]:

$$E_k = -\nabla\Gamma \quad k = \mathrm{x,z} \tag{5.7}$$

where Γ is the electric potential, which is given for top and bottom layers as:

$$\Gamma^t(x,z,t) = -\cos\left(\frac{\pi(z - h_c/2)}{h_t}\right)\varphi^t(x,t) \\ + \frac{2V_0(z - h_c/2)}{h_t}, \tag{5.8}$$

$$\Gamma^b(x,z,t) = -\cos\left(\frac{-\pi(z + h_c/2)}{h_b}\right)\varphi^b(x,t), \tag{5.9}$$

where V_0 is the external voltage that is applied to the actuator layer. The potential strain energy in the structure is given as follows:

$$U = \frac{1}{2}\int_V \left(\sigma_{xx}^c \varepsilon_{xx} + \tau_{xz}^c \gamma_{xz} \right) dV + \frac{1}{2}\int_V \left(\sigma_{xx}^p \varepsilon_{xx} + \tau_{xz}^p \varepsilon_{xz} - D_x E_x - D_z E_z \right) dV, \quad (5.10)$$

By substituting Eqs. (5.1) to (5.6) into Eq. (5.10) we have:

$$
\begin{aligned}
U = &\frac{1}{2}\int_0^L \left[\int \left[N_x \left(\left(\frac{\partial u}{\partial x} \right) \right) - M_x \left(\frac{\partial^2 w}{\partial x^2} \right) + F_x \left(\frac{\partial^2 w}{\partial x^2} - \frac{\partial \psi}{\partial x} \right) + Q_x \left(\left(\frac{\partial w}{\partial x} - \psi \right) \right) \right] dx \right. \\
&+ \int_V \left(\begin{matrix} -D_x \left(\cos\left(\frac{\pi(z - h_c/2)}{h_t} \right) \frac{\partial \phi^t}{\partial x} \right) \\ -D_z \left(-\frac{\pi}{h}\sin\left(\frac{\pi(z - h_c/2)}{h_t} \right) \phi^t - \frac{2V_0}{h} \right) \end{matrix} \right) dV \\
&+ \int_V \left(-D_x \left(\cos\left(\frac{-\pi(z + h_c/2)}{h_b} \right) \frac{\partial \phi^b}{\partial x} \right) - D_z \left(-\frac{\pi}{h}\sin\left(\frac{-\pi}{h_b} \right) \phi^b \right) \right) dV,
\end{aligned}
\qquad (5.11)
$$

where the stress resultants are:

$$N_x = \int \sigma_{xx}^c dA^c + \int \sigma_{xx}^p dA^p, \qquad (5.12)$$

$$M_x = \int \sigma_{xx}^c z \, dA^c + \int \sigma_{xx}^p z \, dA^p, \qquad (5.13)$$

$$F_x = \int \sigma_{xx}^c \Phi(z) dA^c + \int \sigma_{xx}^p \Phi(z) dA^p, \qquad (5.14)$$

$$Q_x = \int \tau_{xz}^c \frac{\partial \Phi(z)}{\partial z} dA^c + \int \tau_{xz}^p \frac{\partial \Phi(z)}{\partial z} dA^p. \qquad (5.15)$$

Substituting Eqs. (5.1)–(5.6) into Eqs. (5.12)–(5.15), the stress resultants can be calculated as:

$$N_x = A_{11}\left(\frac{\partial u}{\partial x} + \frac{1}{2}\left(\frac{\partial w}{\partial x} \right)^2 \right) - B_{11}\left(\frac{\partial^2 w}{\partial x^2} \right) + E_{11}\left(\frac{\partial^2 w}{\partial x^2} - \frac{\partial \psi}{\partial x} \right), \qquad (5.16)$$

$$M_x = B_{11}\left(\frac{\partial u}{\partial x} + \frac{1}{2}\left(\frac{\partial w}{\partial x}\right)^2\right) - D_{11}\left(\frac{\partial^2 w}{\partial x^2}\right) + F_{11}\left(\frac{\partial^2 w}{\partial x^2} - \frac{\partial \psi}{\partial x}\right) + \Xi_{31}\phi, \qquad (5.17)$$

$$F_x = E_{11}\left(\frac{\partial u}{\partial x} + \frac{1}{2}\left(\frac{\partial w}{\partial x}\right)^2\right) - F_{11}\left(\frac{\partial^2 w}{\partial x^2}\right) + H_{11}\left(\frac{\partial^2 w}{\partial x^2} - \frac{\partial \psi}{\partial x}\right) + \Im_{31}\phi, \qquad (5.18)$$

$$Q_x = L_{44}\left(\frac{\partial w}{\partial x} - \psi\right) + \Re_{15}\phi, \qquad (5.19)$$

in which

$$A_{11} = \int C_{11}\,dA^c + \int Q_{11}\,dA^p, \qquad (5.20)$$

$$B_{11} = \int C_{11}\,zdA^c + \int Q_{11}\,zdA^p, \qquad (5.21)$$

$$E_{11} = \int C_{11}\,\Phi(z)dA^c + \int Q_{11}\,\Phi(z)dA^p, \qquad (5.22)$$

$$F_{11} = \int C_{11}\,z\Phi(z)dA^c + \int Q_{11}\,z\Phi(z)dA^p, \qquad (5.23)$$

$$H_{11} = \int C_{11}\,\Phi(z)^2\,dA^c + \int Q_{11}\,\Phi(z)^2\,dA^p, \qquad (5.24)$$

$$L_{44} = \int C_{44}\,\frac{\partial \Phi(z)}{\partial z}\,dA^c + \int Q_{44}\,\frac{\partial \Phi(z)}{\partial z}\,dA^p, \qquad (5.25)$$

$$\Xi_{31} = \int e_{31}\left(\frac{\pi}{h}\sin\left(\frac{\pi z}{h}\right)\right)dA^c, \qquad (5.26)$$

$$\Im_{31} = \int e_{31}\left(\frac{\pi}{h}\sin\left(\frac{\pi z}{h}\right)\right)\Phi(z)dA^c, \qquad (5.27)$$

$$\Re_{15} = -\int e_{15}\left(\cos\left(\frac{\pi z}{h}\right)\right)\frac{\partial \Phi(z)}{\partial z}\,dA^c. \qquad (5.28)$$

The kinetic energy of the structure is defined as follows:

$$K = \frac{\rho}{2}\int\left(\dot{u}_1^2 + \dot{u}_2^2 + \dot{u}_3^2\right)dV, \qquad (5.29)$$

By substituting Eqs. (1.22) to (1.24) into Eq. (5.29), we have:

$$K = \frac{\rho_c + \rho_p}{2} \int \left[\left(\frac{\partial u}{\partial t} - z \frac{\partial^2 w}{\partial x \partial t} + z \left(\frac{\partial^2 w}{\partial x \partial t} - \frac{\partial \psi}{\partial t} \right) \right)^2 + \left(\frac{\partial w}{\partial t} \right)^2 \right] dV, \qquad (5.30)$$

where ρ_c and ρ_p are the density of the core and piezoelectric layers, respectively. By defining the inertia moment terms as:

$$\begin{bmatrix} I_0 \\ I_1 \\ I_2 \\ I_3 \\ I_4 \\ I_5 \end{bmatrix} = \int \begin{bmatrix} \rho_c \\ \rho_c z \\ \rho_c z^2 \\ \rho_c \Phi(z) \\ \rho_c z \Phi(z) \\ \rho_c \Phi(z)^2 \end{bmatrix} dA^c + \int \begin{bmatrix} \rho_p \\ \rho_p z \\ \rho_p z^2 \\ \rho_p \Phi(z) \\ \rho_p z \Phi(z) \\ \rho_p \Phi(z)^2 \end{bmatrix} dA^p, \qquad (5.31)$$

Eq. (5.30) can be rewritten as follows:

$$K = 0.5 \int \left[I_0 \left(\left(\frac{\partial u}{\partial t} \right)^2 + \left(\frac{\partial w}{\partial t} \right)^2 \right) - 2 I_1 \left(\frac{\partial u}{\partial t} \frac{\partial^2 w}{\partial x \partial t} \right) + I_2 \left(\frac{\partial^2 w}{\partial x \partial t} \right)^2 \right.$$

$$\left. + I_3 \frac{\partial u}{\partial t} \left(\frac{\partial^2 w}{\partial x \partial t} - \frac{\partial \psi}{\partial t} \right) - I_4 \frac{\partial^2 w}{\partial x \partial t} \left(\frac{\partial^2 w}{\partial x \partial t} - \frac{\partial \psi}{\partial t} \right) + I_5 \left(\frac{\partial^2 w}{\partial x \partial t} - \frac{\partial \psi}{\partial t} \right)^2 \right] dx. \qquad (5.32)$$

The external work due to the Pasternak foundation can be expressed as:

$$W_k = -\int (-k_w w + k_g \nabla^2 w) w dA, \qquad (5.33)$$

where k_w and k_g are spring constants and shear constants of foundation, respectively. Based on Hamilton's principle, we have:

$$\delta u : \frac{\partial N_x}{\partial x} = I_0 \frac{\partial^2 u}{\partial t^2} + (I_3 - I_1) \frac{\partial^3 w}{\partial x \partial t^2} - I_3 \frac{\partial^2 \psi}{\partial t^2}, \qquad (5.34)$$

$$\delta w : \frac{\partial^2 M_x}{\partial x^2} + 2 e_{31} V_0 \frac{\partial^2 w}{\partial x^2} - \frac{\partial^2 F_x}{\partial x^2} + \frac{\partial Q_x}{\partial x} - k_w w + k_g \nabla^2 w$$

$$= I_0 \frac{\partial^2 w}{\partial t^2} + (I_1 - I_3) \frac{\partial^3 u}{\partial x \partial t^2} + (2 I_4 - I_2 - I_5) \frac{\partial^4 w}{\partial x^2 \partial t^2} + (I_5 - I_4) \frac{\partial^3 \psi}{\partial x \partial t^2}, \qquad (5.35)$$

$$\delta \psi : Q_x - \frac{\partial F_x}{\partial x} = I_5 \frac{\partial^2 \psi}{\partial t^2} - I_3 \frac{\partial^2 u}{\partial t^2} + (I_4 - I_5) \frac{\partial^3 w}{\partial x \partial t^2}, \qquad (5.36)$$

$$\delta\phi^t : \int_{h_c/2}^{h_c/2+h_t} \left(\frac{\partial D_x}{\partial x} \left(\cos\left(\frac{\pi(z-h_c/2)}{h_t}\right) \right) + D_z \left(-\frac{\pi}{h} \sin\left(\frac{\pi(z-h_c/2)}{h_t}\right) \right) \right) dz = 0 \qquad (5.37)$$

$$\delta\phi^b : \int_{-h_c/2}^{-h_c/2-h_b} \left(\frac{\partial D_x}{\partial x} \left(\cos\left(\frac{-\pi(z+h_c/2)}{h_b}\right) \right) + D_z \left(-\frac{\pi}{h} \sin\left(\frac{-\pi}{h_b}\right) \right) \right) dz = 0. \qquad (5.38)$$

Substituting Eqs. (5.16)–(5.19) into Eqs. (5.34)–(5.38), the motion equations of the structure can be derived.

5.3 NUMERICAL RESULTS

In this section, the effects of different parameters based on the DQM presented in Chapter 2 are shown on the DIR of the sandwich structure. The core of the structure is polyethylene with a Young's modulus of $E_c = 1$ GPa integrated with polyvinylidene fluoride (PVDF) with a Young's modulus of $E_p = 1.1$ GPa.

The convergence of the proposed method is shown in Figure 5.2. It can be found that with rising the grid point numbers in the DQM, the DIR shifts to lower frequencies, and finally, at N = 17, the results converge.

For comparison of the results predicted by the hyperbolic theory, the DIR is calculated using the Euler theory, assuming $\Phi(z) = 0$. It can be seen from Figure 5.3 that

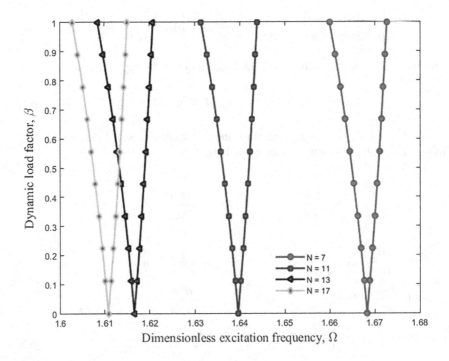

FIGURE 5.2 The convergence of the proposed method.

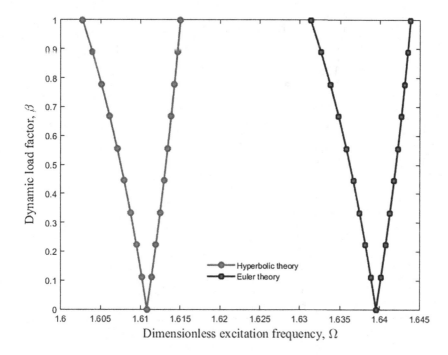

FIGURE 5.3 Comparison of the results predicted by the hyperbolic and Euler theories.

the DIR predicted by hyperbolic theory occurred at a lower excitation frequency than those estimated by the Euler theory. This is because the flexibility of the hyperbolic theory is higher than that of the Euler theory.

Figure 5.4 demonstrates the effect of various boundary conditions on the DIR. Three boundary conditions, including clamped–clamped (CC), clamped–simple (CS), and simple–simple (SS), are considered. It is concluded that the DIR of the beam with the CC boundary condition is observed at a higher excitation frequency than that of the beams with SS and CS boundary conditions. This is because the beam with the CC boundary condition has greater bending rigidity.

The influence of the non-dimensional applied electric voltage $\left(V^* = \left(V_0 / h_t\right)\right.$ $\left.\sqrt{E_c / \in_{11}}\right)$ on the DIR of the sandwich structure is shown in Figure 5.5. It can be found that with applying a positive voltage, the DIR shifts to a lower excitation frequency. This is because of the generation of the tensile and compressive forces resulting from the positive and negative voltage applied to the piezoelectric layer (i.e., actuator), respectively.

The effect of the length-to-total-thickness ratio of the structure on the DIR is shown in Figure 5.6. As can be seen, by raising the length-to-total-thickness ratio of the structure, the DIR shifts to a lower excitation frequency. This is because by raising the length-to-total-thickness ratio of the structure, the stiffness is reduced.

Figure 5.7 depicts the influence of the top-to-core thickness ratio on the DIR of the sandwich structure. It can be concluded that with an increasing top-to-core thickness

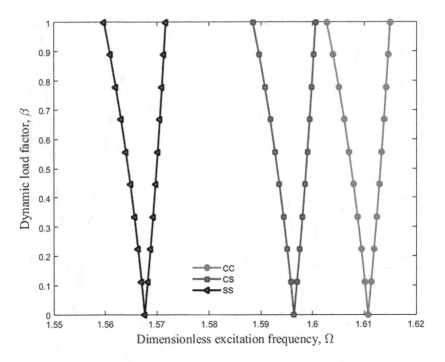

FIGURE 5.4 The effect of different boundary conditions on the DIR of the structure.

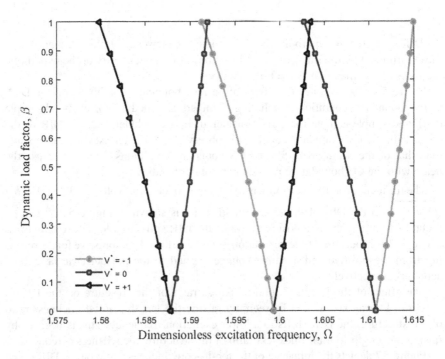

FIGURE 5.5 The effect of the applied electric voltage on the DIR of the structure.

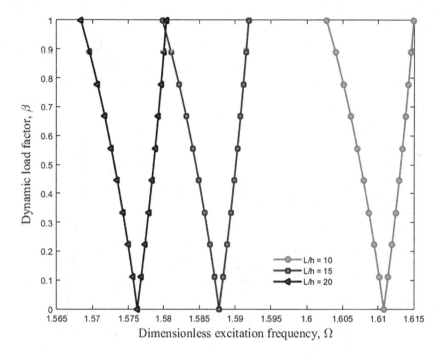

FIGURE 5.6 The effect of length-to-total-thickness ratio on the DIR of the structure.

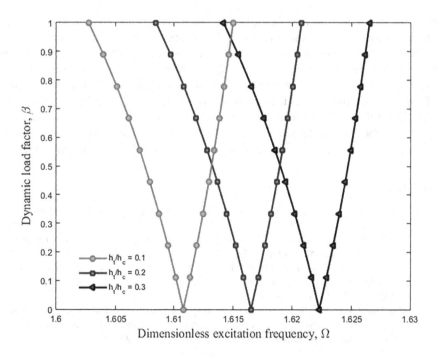

FIGURE 5.7 The effect of top-to-core thickness ratio on the DIR of the structure.

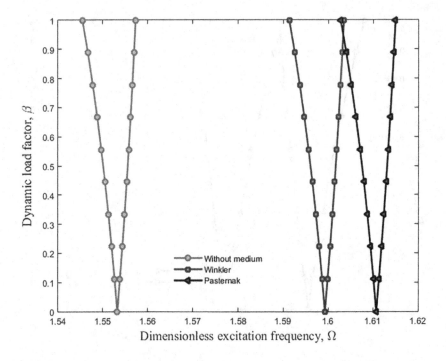

FIGURE 5.8 The effect of the elastic foundation on the DIR of the structure.

ratio, the DIR will be observed at higher excitation frequencies. The reason is that with an increasing top-to-core thickness ratio, the stiffness of the structure is improved.

Figure 5.8 demonstrates the influence of different types of elastic media on the DIR of the structure. As can be seen, considering the elastic medium causes an increase in the excitation frequency of the structure. The reason is that the existence of the elastic medium makes the system stiffer. Furthermore, it can be concluded that the DIR for the Pasternak medium will be observed at higher excitation frequencies than those for the Winkler model. This is because the stiffness of the structure with a Pasternak foundation is higher than that of a structure with the Winkler model.

REFERENCES

1. Yang H, Yu L. Feature extraction of wood-hole defects using wavelet-based ultrasonic testing. *J. Forest Res.* 2017;28:395–402.
2. Henderson JP, Plummer A, Johnston N. An electro-hydrostatic actuator for hybrid active-passive vibration isolation. *Int. J. Hydromech.* 2018;1:47–71.
3. Duc ND, Seung-Eock K, Quan TQ, Long DD, Anh VM. Nonlinear dynamic response and vibration of nanocomposite multilayer organic solar cell. *Compos. Struct.* 2018;184:1137–1144.
4. Van Thu P, Duc ND. Non-linear dynamic response and vibration of an imperfect three-phase laminated nanocomposite cylindrical panel resting on elastic foundations in thermal environment. *Sci. Eng. Compos. Mat.* 2016;24(6):951–962.

5. Shokravi M, Jalili N. Dynamic buckling response of temperature-dependent functionally graded-carbon nanotubes-reinforced sandwich microplates considering structural damping. *Smart. Struct. Syst.* 2017;20:583–593.

6. Shokravi M. Buckling of sandwich plates with FG-CNT-reinforced layers resting on orthotropic elastic medium using Reddy plate theory. *Steel Compos. Struct.* 2017;23:623–631.

7. Shokravi M. Buckling analysis of embedded laminated plates with agglomerated CNT-reinforced composite layers using FSDT and DQM. *Geomech. Eng.* 2017;12:327–346.

8. Duc ND, Hadavinia H, Van Thu P, Quan TQ. Vibration and nonlinear dynamic response of imperfect three-phase polymer nanocomposite panel resting on elastic foundations under hydrodynamic loads. *Compos. Struct.* 2015;131:229–237.

9. Duc ND, Cong PH, Tuan ND, Tran P, Van Thanh N. Thermal and mechanical stability of functionally graded carbon nanotubes (FG CNT)-reinforced composite truncated conical shells surrounded by the elastic foundation. *Thin-Wall Struct.* 2017;115:300–310.

10. Duc ND, Lee J, Nguyen-Thoi T, Thang PT. Static response and free vibration of functionally graded carbon nanotube-reinforced composite rectangular plates resting on Winkler–Pasternak elastic foundations. *Aerosp. Sci. Technol.* 2017;68:391–402.

11. Duc ND, Tran QQ, Nguyen DK. New approach to investigate nonlinear dynamic response and vibration of imperfect functionally graded carbon nanotube reinforced composite double curved shallow shells subjected to blast load and temperature. *Aerosp. Sci. Technol.* 2017;71:360–372.

12. Thanh NV, Khoa ND, Tuan ND, Tran P, Duc ND. Nonlinear dynamic response and vibration of functionally graded carbon nanotube-reinforced composite (FG-CNTRC) shear deformable plates with temperature-dependent material properties. *J. Therm. Stres.* 2017;40:1254–1274.

13. Katariya PV, Panda SK, Hirwani CK, Mehar K, Thakare O. Enhancement of thermal buckling strength of laminated sandwich composite panel structure embedded with shape memory alloy fibre. *Smart Struct. Syst.* 2017;20:595–605.

14. Cafarova FI, Akbarov SD, Yahnioglu N. Buckling delamination of the PZT/Metal/PZT sandwich circular plate-disc with penny-shaped interface cracks. *Smart Struct. Syst.* 2017;19:163–179.

15. Chai YY, Song ZG, Li FM. Investigations on the influences of elastic foundations on the aerothermoelastic flutter and thermal buckling properties of lattice sandwich panels in supersonic airflow. *Acta Astronautic.* 2017;140:176–189.

16. Sayyad AS, Ghugal YM. Bending, buckling and free vibration of laminated composite and sandwich beams: A critical review of literature. *Compos. Struct.* 2017;171:486–504.

17. Song ZG, Li FM. Flutter and buckling characteristics and active control of sandwich panels with triangular lattice core in supersonic airflow. *Compos. Part B: Eng.* 2018;108:334–344.

18. Van Do VN, Lee CH. Thermal buckling analyses of FGM sandwich plates using the improved radial point interpolation mesh-free method. *Compos. Struct.* 2017;177:171–186.

19. Latifi M, Kharazi M, Ovesy HR. Nonlinear dynamic instability analysis of sandwich beams with integral viscoelastic core using different criteria. *Compos. Struct.* 2018;191:89–99.

20. Li X, Yu K, Zhao R. Thermal post-buckling and vibration analysis of a symmetric sandwich beam with clamped and simply supported boundary conditions. *Arch. Appl. Mech.* 2018;88:543–561.

21. Shamshuddin Sayyad A, Ghugal YM. Effect of thickness stretching on the static deformations, natural frequencies, and critical buckling loads of laminated composite and sandwich beams. *J. Braz. Soc. Mech. Sci. Eng.* 2018;40:296–302.

22. Trinh MC, Kim SE. Nonlinear thermomechanical behaviors of thin functionally graded sandwich shells with double curvature. *Compos. Struct.* 2018;195:335–348.
23. Madani H, Hosseini H, Shokravi M. Differential cubature method for vibration analysis of embedded FG-CNT-reinforced piezoelectric cylindrical shells subjected to uniform and non-uniform temperature distributions. *Steel Compos. Struct.* 2016;22:889–913.
24. Shokravi M. Vibration analysis of silica nanoparticles-reinforced concrete beams considering agglomeration effects. *Comput. Concr.* 2017;19:333–338.

6 Buckling of Nanocomposite Plates

6.1 INTRODUCTION

Composite materials are widely used in many branches of industry such as aerospace, civil, naval, and other high-performance engineering applications due to the high stiffness-to-weight ratio. The mechanical behaviors of laminated composite plates are strongly dependent on the degree of orthotropy of individual layers, the low ratio of transverse shear modulus to the in-plane modulus, and the stacking sequence of laminates. A clear understanding of the mechanical analysis of laminated structures is required to achieve the full range of capabilities for the exemplary performance of laminated composite/sandwich structures. Hence, the buckling of laminated plates with nanocomposite layers in order to achieve the high stiffness is presented.

Recent advancements in materials science and computational mechanics have paved the way for significant progress in understanding the behavior of composite structures under various loading conditions. This introduction provides an overview of recent studies in this field, highlighting key contributions and insights from the literature. Chan et al. [1] delved into the axial buckling of multi-walled carbon nanotubes and nanopeapods, providing valuable insights into the buckling behavior of nanostructures. By examining the stability of nanotube-based materials, their study contributes to the understanding of nanoscale mechanics and its implications for various applications. El-Hassar et al. [2] explored the thermal stability analysis of solar functionally graded plates on elastic foundations using an efficient hyperbolic shear deformation theory. Their work addresses the thermal behavior of functionally graded materials, which are increasingly utilized in renewable energy applications, offering valuable insights for design and optimization. Han [3] investigated blood vessel buckling within soft surrounding tissue and its role in generating tortuosity. Understanding the biomechanics of blood vessels is crucial for biomedical applications, and Han's study provides valuable insights into the mechanical behavior of biological structures. Kadoli and Ganesan [4] focused on the free vibration and buckling analysis of composite cylindrical shells conveying hot fluid, providing insights into the dynamic behavior of fluid-conveying structures. Their work contributes to the understanding of fluid–structure interactions and the design of resilient composite materials for engineering applications. Kolahchi et al. [5, 6] explored the dynamic stability and nonlinear dynamic response of temperature-dependent functionally graded carbon nanotube (CNT)-reinforced plates, addressing the challenges associated with the dynamic behavior of advanced composite materials. Their studies provide valuable insights into the mechanical performance of CNT-reinforced composites under varying environmental conditions. Continuing with the exploration of composite materials, Luong-Van et al. [7] developed a cell-based smoothed finite

DOI: 10.1201/9781003510710-6

element method for the dynamic response of laminated composite plates on visco-elastic foundations. By addressing the challenges associated with dynamic analysis, their work contributes to the development of efficient computational techniques for composite structures. Matsunaga [8] investigated the vibration and stability of cross-ply laminated composite plates using a global higher-order plate theory. Their study enhances our understanding of the vibrational characteristics of laminated composites, providing valuable insights for structural design and optimization. Nguyen-Thoi et al. [9–11] developed advanced computational methods for analyzing the dynamic responses of composite plates on viscoelastic foundations, cracked plates, and laminated plates, respectively. By employing innovative numerical techniques, their studies offer new perspectives on the dynamic behavior of composite structures under various loading conditions.

In addition to the aforementioned studies, several other significant contributions have been made by researchers across various domains within composite mechanics. Nguyen et al. [12] focused on refining quasi-3D isogeometric analysis techniques for functionally graded micro plates based on the modified couple stress theory, offering insights into the behavior of microstructures at small scales. Noor [13] investigated the stability of multilayered composite plates, shedding light on the structural integrity of complex composite systems under different loading conditions. Phung-Van and colleagues [14–22] have conducted extensive research on various aspects of composite mechanics, ranging from the development of advanced computational methods for analyzing dynamic responses to the study of thermal and mechanical stability of functionally graded plates. Putcha and Reddy [23] proposed a stiffened plate buckling model for calculating critical stress in the distortional buckling of cold-formed steel beams, contributing to the understanding of structural stability in steel constructions. Saidi et al. [24] introduced a simple hyperbolic shear deformation theory for the vibration analysis of thick functionally graded rectangular plates resting on elastic foundations, providing insights into the behavior of functionally graded materials under dynamic loading conditions. Sheng and Wang [25] investigated the response and control of functionally graded laminated piezoelectric shells under thermal shock and moving loadings, addressing challenges related to the design and optimization of smart composite structures. Tran et al. [26–29] have contributed to various aspects of composite mechanics, including geometrically nonlinear isogeometric analysis, dynamic stability of functionally graded plates, and nonlinear transient analysis of smart piezoelectric plates, offering new perspectives on the behavior of composite structures under complex loading scenarios. Thai et al. [30, 31] have proposed novel theories for the analysis of composite and sandwich plates, providing efficient computational approaches for studying the structural response of advanced composite materials. Zhu and Li [31] developed a stiffened plate buckling model for calculating critical stress in the distortional buckling of cold-formed steel beams, enhancing our understanding of structural stability in steel constructions. Collectively, these research efforts contribute to the advancement of knowledge in composite mechanics and pave the way for innovations in structural design and optimization, with implications for a wide range of engineering applications.

However, buckling analysis of nanocomposite laminated plates has not been performed by researchers. In the present chapter, the orthotropic Mindlin plate theory is used for the nonlinear buckling behavior of embedded laminated plates with nano-composite layers. The layers are reinforced with agglomerated single-walled carbon nanotubes (SWCNTs), and the Mori–Tanaka model is applied for nanocomposite layers. After deriving the governing equations using the energy method and Hamilton's principal, the DQM is applied for determining the buckling load of the structure. The effects of the volume percent of SWCNTs, SWCNT agglomeration, number of layers, orientation angle of layers, boundary conditions, elastic medium, and axial mode number of the plate on the buckling of the structure are disused in detail.

6.2 MOTION EQUATIONS

An embedded laminated plate with nanocomposite layers reinforced with agglom-erated SWCNTs is shown in Figure 6.1. The geometrical parameters of laminated plate are length a, width b, and total thickness h. The elastic medium is simulated by transverse shear loads (k_g) and normal loads (k_w). This chapter is based on Mindlin plate theory, which is presented in Eqs. (1.33) to (1.40) of Chapter 1.

In this chapter, the effective modulus of a laminated plate reinforced by SWCNTs is developed based on the Mori–Tanaka method. The matrix is assumed to be isotro-pic and elastic, with the Young's modulus E_m and the Poisson's ratio v_m. The exper-imental results show that the assumption of uniform dispersion for nanoparticles in the matrix is not correct, and most of the nanoparticles are bent and centralized to one area of the matrix. These regions with concentrated nanoparticles are assumed to have spherical shapes and are considered "inclusions" with different elastic prop-erties from the surrounding material. The total volume V_r of the nanoparticles can be divided into the following two parts [25]:

$$V_r = V_r^{inclusion} + V_r^m, \tag{6.1}$$

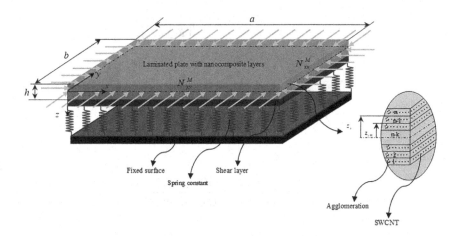

FIGURE 6.1 An embedded laminated plate with agglomerated SWCNT-reinforced layers.

where $V_r^{inclusion}$ and V_r^m are the volumes of nanoparticles dispersed in the spherical inclusions and in the matrix, respectively. The two parameters ξ and ζ describe the agglomeration of nanoparticles:

$$\xi = \frac{V_{inclusion}}{V}, \tag{6.2}$$

$$\zeta = \frac{V_r^{inclusion}}{V_r}. \tag{6.3}$$

However, the average volume fraction C_r of nanoparticles in the composite is:

$$C_r = \frac{V_r}{V}. \tag{6.4}$$

Assuming that all the orientations of the nanoparticles are completely random, the elastic modulus (E) and Poisson's ratio (v) can be calculated as:

$$E = \frac{9KG}{3K+G}, \tag{6.5}$$

$$v = \frac{3K-2G}{6K+2G}, \tag{6.6}$$

where the effective bulk modulus (K) and effective shear modulus (G) are expressed as:

$$K = K_{out}\left[1+\frac{\xi\left(\frac{K_{in}}{K_{out}}-1\right)}{1+\alpha(1-\xi)\left(\frac{K_{in}}{K_{out}}-1\right)}\right], \tag{6.7}$$

$$G = G_{out}\left[1+\frac{\xi\left(\frac{G_{in}}{G_{out}}-1\right)}{1+\beta(1-\xi)\left(\frac{G_{in}}{G_{out}}-1\right)}\right], \tag{6.8}$$

where

$$K_{in} = K_m + \frac{(\delta_r - 3K_m\chi_r)C_r\zeta}{3(\xi - C_r\zeta + C_r\zeta\chi_r)}, \tag{6.9}$$

$$K_{out} = K_m + \frac{C_r(\delta_r - 3K_m\chi_r)(1-\zeta)}{3\left[1-\xi-C_r(1-\zeta)+C_r\chi_r(1-\zeta)\right]}, \tag{6.10}$$

$$G_{in} = G_m + \frac{(\eta_r - 3G_m\beta_r)C_r\zeta}{2(\xi - C_r\zeta + C_r\zeta\beta_r)},$$ (6.11)

$$G_{out} = G_m + \frac{C_r(\eta_r - 3G_m\beta_r)(1-\zeta)}{2[1-\xi-C_r(1-\zeta)+C_r\beta_r(1-\zeta)]},$$ (6.12)

where $\chi_r, \beta_r, \delta_r, \eta_r$ may be calculated as:

$$\chi_r = \frac{3(K_m + G_m) + k_r - l_r}{3(k_r + G_m)},$$ (6.13)

$$\beta_r = \frac{1}{5}\left\{\frac{4G_m + 2k_r + l_r}{3(k_r + G_m)} + \frac{4G_m}{(p_r + G_m)} + \frac{2[G_m(3K_m + G_m) + G_m(3K_m + 7G_m)]}{G_m(3K_m + G_m) + m_r(3K_m + 7G_m)}\right\},$$ (6.14)

$$\delta_r = \frac{1}{3}\left[n_r + 2l_r + \frac{(2k_r - l_r)(3K_m + 2G_m - l_r)}{k_r + G_m}\right],$$ (6.15)

$$\eta_r = \frac{1}{5}\left[\begin{array}{l} \frac{2}{3}(n_r - l_r) + \frac{4G_m p_r}{(p_r + G_m)} + \frac{8G_m m_r(3K_m + 4G_m)}{3K_m(m_r + G_m) + G_m(7m_r + G_m)} \\ + \frac{2(k_r - l_r)(2G_m + l_r)}{3(k_r + G_m)} \end{array}\right],$$ (6.16)

where $k_r, l_r, n_r, p_r,$ and m_r are the Hill's elastic modulus for the nanoparticles. K_m and G_m are the bulk and shear moduli of the matrix, which can be written as:

$$K_m = \frac{E_m}{3(1-2v_m)},$$ (6.17)

$$G_m = \frac{E_m}{2(1+v_m)}.$$ (6.18)

Furthermore, β, α can be obtained from:

$$\alpha = \frac{(1+v_{out})}{3(1-v_{out})},$$ (6.19)

$$\beta = \frac{2(4 - 5v_{out})}{15(1-v_{out})},$$ (6.20)

$$v_{out} = \frac{3K_{out} - 2G_{out}}{6K_{out} + 2G_{out}}. \tag{6.21}$$

However, using Eqs. (6.5) and (6.6), the elastic constants of the structure can be calculated, which is introduced in the next section. The constitutive equation for the stress (σ) and strain (ε) matrix of the kth layer can be expressed as follows [22]:

$$\begin{Bmatrix} \sigma_{xx} \\ \sigma_{yy} \\ \sigma_{xy} \\ \sigma_{xz} \\ \sigma_{yz} \end{Bmatrix}^{(k)} = \begin{bmatrix} Q_{11} & Q_{12} & Q_{16} & 0 & 0 \\ Q_{12} & Q_{22} & Q_{26} & 0 & 0 \\ Q_{16} & Q_{26} & Q_{66} & 0 & 0 \\ 0 & 0 & 0 & Q_{55} & Q_{45} \\ 0 & 0 & 0 & Q_{45} & Q_{44} \end{bmatrix}^{(k)} \begin{Bmatrix} \varepsilon_{xx} \\ \varepsilon_{yy} \\ \gamma_{xy} \\ \gamma_{xz} \\ \gamma_{yz} \end{Bmatrix}^{(k)}, \tag{6.22}$$

where Q_{ij} $(i, j = 1, 2, ..., 6)$ is defined as:

$$Q_{11} = C_{11} \cos^4 \theta - 4C_{16} \cos^3 \theta \sin \theta + 2(C_{12} + 2C_{66}) \cos^2 \theta \sin^2 \theta$$
$$- 4C_{26} \cos \theta \sin^3 \theta + C_{22} \sin^4 \theta, \tag{6.23}$$

$$Q_{12} = C_{12} \cos^4 \theta + 2(C_{16} - C_{26}) \cos^3 \theta \sin \theta + (C_{11} + C_{22} - 4C_{66}) \cos^2 \theta \sin^2 \theta$$
$$+ 2(C_{26} - C_{16}) \cos \theta \sin^3 \theta + C_{12} \sin^4 \theta, \tag{6.24}$$

$$Q_{16} = C_{16} \cos^4 \theta + (C_{11} - C_{12} - 2C_{66}) \cos^3 \theta \sin \theta + 3(C_{26} - C_{16}) \cos^2 \theta \sin^2 \theta$$
$$+ (2C_{66} + C_{12} - C_{22}) \cos \theta \sin^3 \theta - C_{26} \sin^4 \theta, \tag{6.25}$$

$$Q_{22} = C_{22} \cos^4 \theta + 4C_{26} \cos^3 \theta \sin \theta + 2(C_{12} + 2C_{66}) \cos^2 \theta \sin^2 \theta$$
$$+ 4C_{16} \cos \theta \sin^3 \theta + C_{11} \sin^4 \theta, \tag{6.26}$$

$$Q_{26} = C_{26} \cos^4 \theta + (C_{12} - C_{22} + 2C_{66}) \cos^3 \theta \sin \theta + 3(C_{16} - C_{26}) \cos^2 \theta \sin^2 \theta$$
$$+ (C_{11} - C_{12} - 2C_{66}) \cos \theta \sin^3 \theta - C_{16} \sin^4 \theta, \tag{6.27}$$

$$Q_{66} = 2(C_{16} - C_{26}) \cos^3 \theta \sin \theta + (C_{11} + C_{22} - 2C_{12} - 2C_{66}) \cos^2 \theta \sin^2 \theta$$
$$+ 2(C_{26} - C_{26}) \cos \theta \sin^3 \theta + C_{66} (\cos^4 \theta + \sin^4 \theta), \tag{6.28}$$

$$Q_{44} = C_{44} \cos^2 \theta + 2C_{45} \cos \theta \sin \theta + C_{55} \sin^2 \theta, \tag{6.29}$$

$$Q_{45} = \left(C_{55} - C_{44}\right)\cos\theta\sin\theta + C_{44}\left(\cos^2\theta - \sin^2\theta\right), \tag{6.30}$$

$$Q_{55} = C_{55}\cos^2\theta - 2C_{45}\cos\theta\sin\theta + C_{44}\sin^2\theta. \tag{6.31}$$

The strain energy U of the structure can be written as:

$$U = \frac{1}{2}\int_V \left(\sigma_{xx}^{(k)}\varepsilon_{xx} + \sigma_{yy}^{(k)}\varepsilon_{yy} + \sigma_{xy}^{(k)}\gamma_{xy} + \sigma_{xz}^{(k)}\gamma_{xz} + \sigma_{yz}^{(k)}\gamma_{yz}\right)dV. \tag{6.32}$$

Combining Eqs. (1.36) to (1.40) and Eq. (6.32) yields:

$$
\begin{aligned}
U = \frac{1}{2}\int_{\Omega_0} &\left[N_{xx}\left(\frac{\partial u}{\partial x} + \frac{1}{2}\left(\frac{\partial w}{\partial x}\right)^2\right) + N_{yy}\left(\frac{\partial v}{\partial y} + \frac{1}{2}\left(\frac{\partial w}{\partial y}\right)^2\right) + Q_y\left(\frac{\partial w_0}{\partial y} + \psi_y\right) \right. \\
&+ Q_x\left(\frac{\partial w_0}{\partial x} + \psi_x\right) + N_{xy}\left(\frac{\partial v}{\partial y} + \frac{\partial u}{\partial x} + \frac{\partial w}{\partial x}\frac{\partial w}{\partial y}\right) + M_{xx}\frac{\partial \psi_x}{\partial x} \\
&\left. + M_{yy}\frac{\partial \psi_y}{\partial x} + M_{xy}\left(\frac{\partial \psi_x}{\partial y} + \frac{\partial \psi_y}{\partial x}\right) \right] dx dy,
\end{aligned}
\tag{6.33}
$$

where the stress resultant–displacement relations can be defined as:

$$\left\{(N_{xx}, N_{yy}, N_{xy}), (M_{xx}, M_{yy}, M_{xy})\right\} = \sum_{k=1}^{N}\int_{z^{(k-1)}}^{z^{(k)}}\left\{\sigma_{xx}, \sigma_{yy}, \tau_{xy}\right\}(1, z)dz, \tag{6.34}$$

$$\left\{Q_x, Q_y\right\} = K\sum_{k=1}^{N}\int_{z^{(k-1)}}^{z^{(k)}}\left\{\sigma_{xz}, \sigma_{yz}\right\}dz, \tag{6.35}$$

in which K is the shear correction coefficient. The external work due to the surrounding elastic medium can be written as:

$$W = -\int\left(k_w w - k_g \nabla^2\right)w dA. \tag{6.36}$$

The governing equations can be derived by Hamilton's principal ($\int_0^t(-\delta U + \delta W)dt = 0$) as follows:

$$\delta u : \frac{\partial N_{xx}}{\partial x} + \frac{\partial N_{xy}}{\partial y} = 0, \tag{6.37}$$

$$\delta v : \frac{\partial N_{xy}}{\partial x} + \frac{\partial N_{yy}}{\partial y} = 0, \tag{6.38}$$

$$\delta w: \frac{\partial Q_x}{\partial x} + \frac{\partial Q_y}{\partial y} + \frac{\partial}{\partial x}\left(N_{xx}^M \frac{\partial w}{\partial x}\right) + \frac{\partial}{\partial y}\left(N_{yy}^M \frac{\partial w}{\partial y}\right) - k_w w + k_g \nabla^2 w = 0, \quad (6.39)$$

$$\delta \psi_x : \frac{\partial M_{xx}}{\partial x} + \frac{\partial M_{xy}}{\partial y} - Q_x = 0, \tag{6.40}$$

$$\delta \psi_y : \frac{\partial M_{xy}}{\partial x} + \frac{\partial M_{yy}}{\partial y} - Q_y = 0, \tag{6.41}$$

where N_{xx}^M and N_{yy}^M are the mechanical forces. Substituting Eq. (6.22) into Eqs. (6.34) and (6.35), the stress resultant–displacement relations can be obtained as follows:

$$N_{xx} = A_{11}\left(\frac{\partial u}{\partial x} + \frac{1}{2}\left(\frac{\partial w}{\partial x}\right)^2\right) + A_{12}\left(\frac{\partial v}{\partial y} + \frac{1}{2}\left(\frac{\partial w}{\partial y}\right)^2\right) + A_{16}\left(\frac{\partial u}{\partial y} + \frac{\partial v}{\partial x} + \frac{\partial w}{\partial x}\frac{\partial w}{\partial y}\right)$$
$$+ B_{11}\frac{\partial \psi_x}{\partial x} + B_{12}\frac{\partial \psi_y}{\partial y} + B_{16}\left(\frac{\partial \psi_x}{\partial y} + \frac{\partial \psi_y}{\partial x}\right),$$

$$(6.42)$$

$$N_{yy} = A_{12}\left(\frac{\partial u}{\partial x} + \frac{1}{2}\left(\frac{\partial w}{\partial x}\right)^2\right) + A_{22}\left(\frac{\partial v}{\partial y} + \frac{1}{2}\left(\frac{\partial w}{\partial y}\right)^2\right) + A_{26}\left(\frac{\partial u}{\partial y} + \frac{\partial v}{\partial x} + \frac{\partial w}{\partial x}\frac{\partial w}{\partial y}\right)$$
$$+ B_{12}\frac{\partial \psi_x}{\partial x} + B_{22}\frac{\partial \psi_y}{\partial y} + B_{26}\left(\frac{\partial \psi_x}{\partial y} + \frac{\partial \psi_y}{\partial x}\right),$$

$$(6.43)$$

$$N_{xy} = A_{16}\left(\frac{\partial u}{\partial x} + \frac{1}{2}\left(\frac{\partial w}{\partial x}\right)^2\right) + A_{26}\left(\frac{\partial v}{\partial y} + \frac{1}{2}\left(\frac{\partial w}{\partial y}\right)^2\right) + A_{66}\left(\frac{\partial u}{\partial y} + \frac{\partial v}{\partial x} + \frac{\partial w}{\partial x}\frac{\partial w}{\partial y}\right)$$
$$+ B_{16}\frac{\partial \psi_x}{\partial x} + B_{26}\frac{\partial \psi_y}{\partial y} + B_{66}\left(\frac{\partial \psi_x}{\partial y} + \frac{\partial \psi_y}{\partial x}\right),$$

$$(6.44)$$

$$M_{xx} = B_{11}\left(\frac{\partial u}{\partial x} + \frac{1}{2}\left(\frac{\partial w}{\partial x}\right)^2\right) + B_{12}\left(\frac{\partial v}{\partial y} + \frac{1}{2}\left(\frac{\partial w}{\partial y}\right)^2\right) + B_{16}\left(\frac{\partial u}{\partial y} + \frac{\partial v}{\partial x} + \frac{\partial w}{\partial x}\frac{\partial w}{\partial y}\right)$$
$$+ D_{11}\frac{\partial \psi_x}{\partial x} + D_{12}\frac{\partial \psi_y}{\partial y} + D_{16}\left(\frac{\partial \psi_x}{\partial y} + \frac{\partial \psi_y}{\partial x}\right),$$

$$(6.45)$$

$$M_{yy} = B_{12}\left(\frac{\partial u}{\partial x} + \frac{1}{2}\left(\frac{\partial w}{\partial x}\right)^2\right) + B_{22}\left(\frac{\partial v}{\partial y} + \frac{1}{2}\left(\frac{\partial w}{\partial y}\right)^2\right) + B_{26}\left(\frac{\partial u}{\partial y} + \frac{\partial v}{\partial x} + \frac{\partial w}{\partial x}\frac{\partial w}{\partial y}\right)$$
$$+ D_{12}\frac{\partial \psi_x}{\partial x} + D_{22}\frac{\partial \psi_y}{\partial y} + D_{26}\left(\frac{\partial \psi_x}{\partial y} + \frac{\partial \psi_y}{\partial x}\right),$$

$$(6.46)$$

$$M_{xy} = B_{16}\left(\frac{\partial u}{\partial x} + \frac{1}{2}\left(\frac{\partial w}{\partial x}\right)^2\right) + B_{26}\left(\frac{\partial v}{\partial y} + \frac{1}{2}\left(\frac{\partial w}{\partial y}\right)^2\right) + B_{66}\left(\frac{\partial u}{\partial y} + \frac{\partial v}{\partial x} + \frac{\partial w}{\partial x}\frac{\partial w}{\partial y}\right)$$
$$+ D_{16}\frac{\partial \psi_x}{\partial x} + D_{26}\frac{\partial \psi_y}{\partial y} + D_{66}\left(\frac{\partial \psi_x}{\partial y} + \frac{\partial \psi_y}{\partial x}\right),$$

$$(6.47)$$

$$Q_{xx} = A_{55}\left(\frac{\partial w}{\partial x} + \psi_x\right) + A_{45}\left(\frac{\partial w}{\partial y} + \psi_y\right),$$

$$(6.48)$$

$$Q_{yy} = A_{45}\left(\frac{\partial w}{\partial y} + \psi_y\right) + A_{44}\left(\frac{\partial w}{\partial y} + \psi_y\right),$$

$$(6.49)$$

where

$$A_{ij} = \sum_{k=1}^{N}\int_{z^{(k-1)}}^{z^{(k)}} Q_{ij}^{(k)}dz, \qquad (i,j=1,2,6) \qquad (6.50)$$

$$B_{ij} = \sum_{k=1}^{N}\int_{z^{(k-1)}}^{z^{(k)}} Q_{ij}^{(k)}z dz, \qquad (6.51)$$

$$D_{ij} = \sum_{k=1}^{N}\int_{z^{(k-1)}}^{z^{(k)}} Q_{ij}^{(k)}z^2 dz. \qquad (6.52)$$

Substituting Eqs. (6.42)–(6.49) into Eqs. (6.37)–(6.41) yields the governing equations as:

$$A_{11}\left(\frac{\partial^2 u}{\partial x^2} + \frac{\partial w}{\partial x}\frac{\partial^2 w}{\partial x^2}\right) + A_{12}\left(\frac{\partial^2 v}{\partial x\partial y} + \frac{\partial w}{\partial y}\frac{\partial^2 w}{\partial x\partial y}\right) + A_{16}\left(\frac{\partial^2 u}{\partial x\partial y} + \frac{\partial^2 v}{\partial x^2} + \frac{\partial w}{\partial y}\frac{\partial^2 w}{\partial x^2} + \frac{\partial w}{\partial x}\frac{\partial^2 w}{\partial x\partial y}\right)$$
$$+ B_{11}\frac{\partial^2 \psi_x}{\partial x^2} + B_{12}\frac{\partial^2 \psi_y}{\partial x\partial y} + B_{16}\left(\frac{\partial^2 \psi_x}{\partial x\partial y} + \frac{\partial^2 \psi_y}{\partial x^2}\right) + A_{16}\left(\frac{\partial^2 u}{\partial x\partial y} + \frac{\partial w}{\partial x}\frac{\partial^2 w}{\partial x\partial y}\right)$$
$$+ A_{26}\left(\frac{\partial^2 v}{\partial y^2} + \frac{\partial w}{\partial y}\frac{\partial^2 w}{\partial y^2}\right) + A_{66}\left(\frac{\partial^2 u}{\partial y^2} + \frac{\partial^2 v}{\partial x\partial y} + \frac{\partial w}{\partial x}\frac{\partial^2 w}{\partial y^2} + \frac{\partial w}{\partial y}\frac{\partial^2 w}{\partial x\partial y}\right) + B_{16}\frac{\partial^2 \psi_x}{\partial x\partial y}$$
$$+ B_{26}\frac{\partial^2 \psi_y}{\partial y^2} + B_{66}\left(\frac{\partial^2 \psi_x}{\partial y^2} + \frac{\partial^2 \psi_y}{\partial x\partial y}\right) = 0,$$

$$(6.53)$$

$$A_{16}\left(\frac{\partial^2 u}{\partial x^2}+\frac{\partial w}{\partial x}\frac{\partial^2 w}{\partial x^2}\right)+A_{26}\left(\frac{\partial^2 v}{\partial x\partial y}+\frac{\partial w}{\partial y}\frac{\partial^2 w}{\partial x\partial y}\right)+A_{66}\left(\frac{\partial^2 u}{\partial x\partial y}+\frac{\partial^2 v}{\partial x^2}+\frac{\partial w}{\partial y}\frac{\partial^2 w}{\partial x^2}+\frac{\partial w}{\partial x}\frac{\partial^2 w}{\partial x\partial y}\right)$$

$$+B_{16}\frac{\partial^2 \psi_x}{\partial x^2}+B_{26}\frac{\partial^2 \psi_y}{\partial x\partial y}+B_{66}\left(\frac{\partial^2 \psi_x}{\partial x\partial y}+\frac{\partial^2 \psi_y}{\partial x^2}\right)+A_{21}\left(\frac{\partial^2 u}{\partial x\partial y}+\frac{\partial w}{\partial x}\frac{\partial^2 w}{\partial x\partial y}\right)$$

$$+A_{22}\left(\frac{\partial^2 v}{\partial y^2}+\frac{\partial w}{\partial y}\frac{\partial^2 w}{\partial y^2}\right)+A_{26}\left(\frac{\partial^2 u}{\partial y^2}+\frac{\partial^2 v}{\partial x\partial y}+\frac{\partial w}{\partial x}\frac{\partial^2 w}{\partial y^2}+\frac{\partial w}{\partial y}\frac{\partial^2 w}{\partial x\partial y}\right)$$

$$+B_{21}\frac{\partial^2 \psi_x}{\partial x\partial y}+B_{22}\frac{\partial^2 \psi_y}{\partial y^2}+B_{26}\left(\frac{\partial^2 \psi_x}{\partial y^2}+\frac{\partial^2 \psi_y}{\partial x\partial y}\right)=0,$$

$$(6.54)$$

$$A_{55}\left(\frac{\partial^2 w}{\partial x^2}+\frac{\partial \psi_x}{\partial x}\right)+A_{45}\left(\frac{\partial^2 w}{\partial x\partial y}+\frac{\partial \psi_y}{\partial x}\right)+A_{45}\left(\frac{\partial^2 w}{\partial x\partial y}+\frac{\partial \psi_x}{\partial y}\right)$$

$$(6.55)$$

$$+A_{44}\left(\frac{\partial^2 w}{\partial y^2}+\frac{\partial \psi_y}{\partial y}\right)+N_{xx}^M\frac{\partial^2 w}{\partial x^2}+N_{yy}^M\frac{\partial^2 w}{\partial x^2}-k_w w+k_g\left(\frac{\partial^2 w}{\partial x^2}+\frac{\partial^2 w}{\partial y^2}\right)=0,$$

$$B_{11}\left(\frac{\partial^2 u}{\partial x^2}+\frac{\partial w}{\partial x}\frac{\partial^2 w}{\partial x^2}\right)+B_{12}\left(\frac{\partial^2 v}{\partial x\partial y}+\frac{\partial w}{\partial y}\frac{\partial^2 w}{\partial x\partial y}\right)+B_{16}\left(\frac{\partial^2 u}{\partial x\partial y}+\frac{\partial^2 v}{\partial x^2}+\frac{\partial w}{\partial y}\frac{\partial^2 w}{\partial x^2}+\frac{\partial w}{\partial x}\frac{\partial^2 w}{\partial x\partial y}\right)$$

$$+D_{11}\frac{\partial^2 \psi_x}{\partial x^2}+D_{12}\frac{\partial^2 \psi_y}{\partial x\partial y}+D_{16}\left(\frac{\partial^2 \psi_x}{\partial x\partial y}+\frac{\partial^2 \psi_y}{\partial x^2}\right)+B_{16}\left(\frac{\partial^2 u}{\partial x\partial y}+\frac{\partial w}{\partial x}\frac{\partial^2 w}{\partial x\partial y}\right)$$

$$+B_{26}\left(\frac{\partial^2 v}{\partial y^2}+\frac{\partial w}{\partial y}\frac{\partial^2 w}{\partial y^2}\right)+B_{66}\left(\frac{\partial^2 u}{\partial y^2}+\frac{\partial^2 v}{\partial x\partial y}+\frac{\partial w}{\partial x}\frac{\partial^2 w}{\partial y^2}+\frac{\partial w}{\partial y}\frac{\partial^2 w}{\partial x\partial y}\right)$$

$$+D_{16}\frac{\partial^2 \psi_x}{\partial x\partial y}+D_{26}\frac{\partial^2 \psi_y}{\partial y^2}+D_{66}\left(\frac{\partial^2 \psi_x}{\partial y^2}+\frac{\partial^2 \psi_y}{\partial x\partial y}\right)-A_{55}\left(\frac{\partial w}{\partial x}+\psi_x\right)-A_{45}\left(\frac{\partial w}{\partial y}+\psi_y\right)=0,$$

$$(6.56)$$

$$B_{16}\left(\frac{\partial^2 u}{\partial x^2}+\frac{\partial w}{\partial x}\frac{\partial^2 w}{\partial x^2}\right)+B_{26}\left(\frac{\partial^2 v}{\partial x\partial y}+\frac{\partial w}{\partial y}\frac{\partial^2 w}{\partial x\partial y}\right)+B_{66}\left(\frac{\partial^2 u}{\partial x\partial y}+\frac{\partial^2 v}{\partial x^2}+\frac{\partial w}{\partial y}\frac{\partial^2 w}{\partial x^2}+\frac{\partial w}{\partial x}\frac{\partial^2 w}{\partial x\partial y}\right)$$

$$+D_{16}\frac{\partial^2 \psi_x}{\partial x^2}+D_{26}\frac{\partial^2 \psi_y}{\partial x\partial y}+D_{66}\left(\frac{\partial^2 \psi_x}{\partial x\partial y}+\frac{\partial^2 \psi_y}{\partial x^2}\right)+B_{21}\left(\frac{\partial^2 u}{\partial x\partial y}+\frac{\partial w}{\partial x}\frac{\partial^2 w}{\partial x\partial y}\right)$$

$$+B_{22}\left(\frac{\partial^2 v}{\partial y^2}+\frac{\partial w}{\partial y}\frac{\partial^2 w}{\partial y^2}\right)+B_{26}\left(\frac{\partial^2 u}{\partial y^2}+\frac{\partial^2 v}{\partial x\partial y}+\frac{\partial w}{\partial x}\frac{\partial^2 w}{\partial y^2}+\frac{\partial w}{\partial y}\frac{\partial^2 w}{\partial x\partial y}\right)$$

$$+D_{21}\frac{\partial^2 \psi_x}{\partial x\partial y}+D_{22}\frac{\partial^2 \psi_y}{\partial y^2}+D_{26}\left(\frac{\partial^2 \psi_x}{\partial y^2}+\frac{\partial^2 \psi_y}{\partial x\partial y}\right)-A_{45}\left(\frac{\partial w}{\partial y}+\psi_y\right)-A_{44}\left(\frac{\partial w}{\partial y}+\psi_y\right)=0.$$

$$(6.57)$$

Based on the DQM presented in Chapter 2, the governing equations can be expressed in matrix form as:

$$\left(\left[K_L + K_{NL}\right] + P\left[K_g\right]\right)\{d\} = 0, \tag{6.58}$$

where $\left[K_G\right]$ is the coefficient force matrix, $\left[K_L\right]$ is the linear stiffness matrix, $\left[K_{NL}\right]$ is the nonlinear stiffness matrix, and $\{d\} = \{u, v, w, \psi_x, \psi_y\}$ is the displacement vector. This nonlinear equation can now be solved using a direct iterative process as follows:

- First, nonlinearity is ignored by taking $K_{NL} = 0$ to solve the eigenvalue problem expressed in Eq. (6.58). This yields the linear eigenvalue (P_L) and associated eigenvector (d). The eigenvector is then scaled up so that the maximum transverse displacement of the structure is equal to the maximum eigenvector, that is, the given vibration amplitude d_{max}.
- Using linear d, $\left[K_{NL}\right]$ can be evaluated. The eigenvalue problem is then solved by substituting $\left[K_{NL}\right]$ into Eq. (6.58). This would give the nonlinear eigenvalue (P_{NL}) and the new eigenvector.
- The new nonlinear eigenvector is scaled up again, and the procedure is repeated iteratively until the buckling load values from the two subsequent iterations, "r" and

"$r+1$" satisfy the prescribed convergence criteria as:

$$\frac{\left|\omega^{r+1} - \omega^r\right|}{\omega^r} < \varepsilon_c, \tag{6.59}$$

where ε_0 is a small value number, and in the present analysis, it is taken to be 0.01%.

6.3 NUMERICAL RESULTS

In this section, a laminated plate with the material properties listed in Table 6.1 is considered.

TABLE 6.1
Material Properties of Graphite/Epoxy [19]

Properties	Value
E_{11}	132.38 GPa
$E_{22} = E_{33}$	10.76 GPa
G_{12}	3.61 GPa
$G_{13} = G_{23}$	5.65 GPa
$\nu_{11} = \nu_{23}$	0.24
ν_{13}	0.49

First, the convergence and accuracy of the proposed method are studied. Then, the results are validated with other published works, and finally, the effects of different parameters such as the number of lamina, orientation angle of the lamina, volume percent of SWCNTs, SWCNT agglomeration, boundary conditions, elastic medium, and axial mode number of the plate are shown on the dimensionless buckling load $(DBL = P/(E_{11}a))$. Three kinds of boundary conditions exist: all edges simply supported (SSSS), all edged clamped (CCCC), and two opposite edges simply supported and the other two clamped (SCSC) (Kolahchi et al. 2016a)

In order to validate the present results, we neglected the agglomerated SWCNTs as reinforcement $(C_r = \xi = \zeta = 0)$, elastic medium constants $(k_w = k_g = 0)$, and nonlinear terms in the governing equations. However, the buckling of a simply supported laminated plate is studied based on the FSDT and DQM considering the material properties, as reported by Matsunaga [8]. The results of comparison are shown in Table 6.2. As can be seen, the present results are in good agreement with Noor [13] based on the 3D elasticity solution, Putcha and Reddy [21]based on the FSDT, and Matsunaga [13] based on the HSDT. Note that the slight difference between the results of this chapter and other works is due to the difference in the theory and solution method.

In all of the following figures, the dimensionless buckling load is plotted against the axial mode number. It can be observed that the buckling load decreases at first until reaching the lowest amount, and after that, the increasing process begins. The critical buckling load appears at the point where the buckling load is minimal.

The effect of the SWCNT volume percent on the dimensionless buckling load versus the axial mode number is shown in Figure 6.2. It can be seen that with an increasing SWCNT volume percent, the dimensionless buckling load increases and

TABLE 6.2
Comparison of the Present Work with this chapter

No. of layers	Solution	E_1/E_2				
		3	10	20	30	40
3	A	5.3044	9.7621	15.0191	19.3040	22.8807
	B	5.3991	9.9652	15.3510	19.7560	23.4530
	C	5.3208	9.7172	14.7290	18.6834	21.8977
	D	5.3918	9.8452	14.9167	18.8769	22.1531
5	A	5.3255	9.9603	15.6527	20.4663	24.5929
	B	5.4093	10.1360	15.9560	20.9080	25.1850
	C	5.3348	9.9414	15.5142	20.1656	24.1158
	D	3.4011	9.9087	15.2451	20.3483	24.9098

A: 3D elasticity solution [13]
B: FSDT [21]
C: HSDT [8]
D: Present work

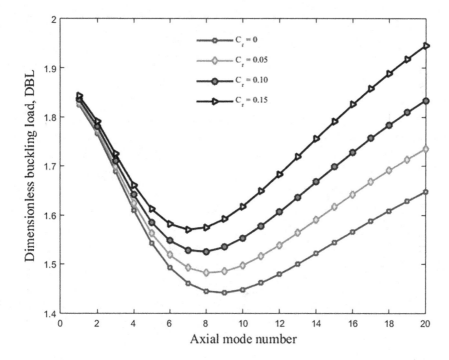

FIGURE 6.2 The effect of SWCNT volume percent on the dimensionless buckling load versus axial mode number.

the critical buckling load is increased. This is due to the fact that with an increasing SWCNT volume percent, the stiffness of the structure increases. In addition, the effect of the SWCNT volume percent on the dimensionless buckling load becomes more prominent at higher axial mode numbers.

In order to show the effects of SWCNT agglomeration on the dimensionless buckling load, Figure 6.3 is plotted. As can be seen, considering agglomeration effects leads to a decrease in the dimensionless buckling load of the structure. This is because the SWCNT agglomeration is a harmful parameter for the system due to a reduction in the stability and rigidity of the structure. However, since the dispersion of SWCNTs in the matrix cannot be uniform in reality, the results of this figure can be useful for the research and design of laminated plates with nanocomposite layers.

Figure 6.4 illustrates the effect of the layer number on the dimensionless buckling load versus axial mode numbers. It can be concluded that the symmetric structure with three layers predicted a higher dimensionless buckling load for all axial mode numbers with respect to the anti-symmetric one with two layers. This is due to the fact that the stability and balance of a structure with symmetric laminas are higher than those of a structure with anti-symmetric laminas.

The effect of the orientation angle of the SWCNTs in the laminas on the dimensionless buckling load versus axial mode number is presented in Figure 6.5. Here, three cases of zero lamina $(0°,0°,0°)$, cross-ply lamina $(0°,90°,0°)$, and angle-ply

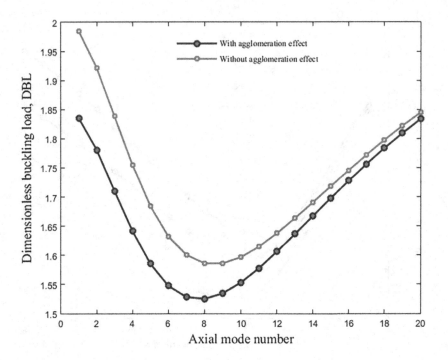

FIGURE 6.3 The effect of SWCNT agglomeration on the dimensionless buckling load versus axial mode number.

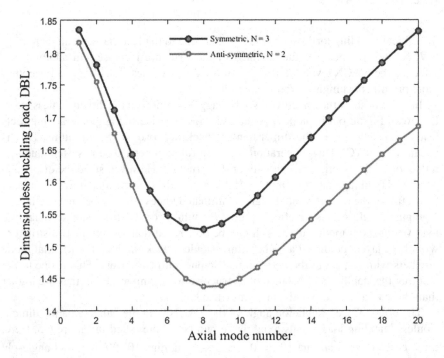

FIGURE 6.4 The effect of lamina numbers on the dimensionless buckling load versus axial mode number.

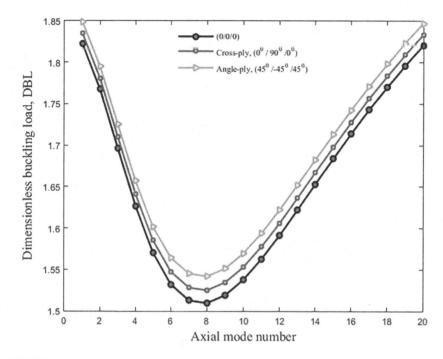

FIGURE 6.5 The effect of the orientation angle of layers on the dimensionless buckling load versus axial mode number.

lamina $\left(45°,-45°,45°\right)$ are assumed. As can be seen, the dimensionless buckling load for the angle-ply lamina is higher than that for cross-ply and zero laminas. In other words, the zero lamina predicts the lowest dimensionless buckling load with respect to other cases.

Figure 6.6 demonstrates the effect of boundary conditions on the dimensionless buckling load versus axial mode number. Three boundary conditions of SSSS, SCSC, and CCCC are considered. It is obvious that the dimensionless buckling load has the following order for the proposed boundary conditions:

CCCC>SCSC>SSSS

Hence, the laminated plate with the CCCC boundaries has a higher dimensionless buckling load than the other considered cases. This is due to the fact that the stiffness of the structure increases for the CCCC boundary condition.

The effect of the elastic medium on the dimensionless buckling load versus axial mode number is shown in Figure 6.7. Generally, the existence of the elastic medium increases the stiffness of the structure, and therefore, the dimensionless buckling load increases. The Pasternak medium considers the vertical and shear loads; however the Winkler medium considers only the vertical ones. Therefore, the effect of Pasternak medium is greater than that of Winkler medium. According to Figure 6.7, the effect of the elastic medium on the buckling load is significant, and it can be a useful parameter to take the system away from the buckling condition.

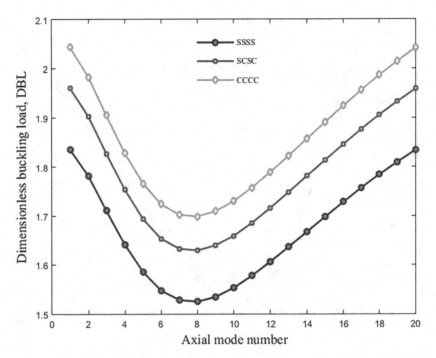

FIGURE 6.6 The effect of boundary conditions on the dimensionless buckling load versus axial mode number.

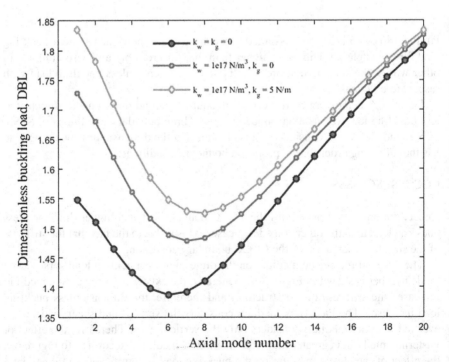

FIGURE 6.7 The effect of an elastic medium on the dimensionless buckling load versus axial mode number.

REFERENCES

1. Chan Y, Thamwattana N, Hill JM. Axial buckling of multi walled carbon nanotubes and nanopeapods. *Europ. J. Mech. A/Solids.* 2011;30:794–806.
2. El-Hassar SM, Benyoucef S, Heireche H, Tounsi A. Thermal stability analysis of solar functionally graded plates on elastic foundation using an efficient hyperbolic shear deformation theory. *Geomech. Eng.* 2016;10:357–386.
3. Han HC. Blood vessel buckling within soft surrounding tissue generates tortuosity. *J. Biomech.* 2009;42:2797–2801.
4. Kadoli R, Ganesan N. Free vibration and buckling analysis of composite cylindrical shells conveying hot fluid. *Compos. Struct.* 2003;60:19–32.
5. Kolahchi R, Safari M, Esmailpour M. Dynamic stability analysis of temperature-dependent functionally graded CNT-reinforced visco-plates resting on orthotropic elastomeric medium. *Compos. Struct.* 2016;150:255–265.
6. Kolahchi R, Hosseini H, Esmailpour M. Differential cubature and quadrature-Bolotin methods for dynamic stability of embedded piezoelectric nanoplates based on visco-nonlocal-piezoelasticity theories. *Compos. Struct.* 2016;157:174–186.
7. Luong-Van H, Nguyen-Thoi T, Liu GR, Phung-Van P. A cell-based smoothed finite element method using three-node shear-locking free Mindlin plate element (CS-FEM-MIN3) for dynamic response of laminated composite plates on viscoelastic foundation. *Eng. Anal. Bound. Elem.* 2014;42:8–19.
8. Matsunaga H. Vibration and stability of cross-ply laminated composite plates according to a global higher-order plate theory. *Compos. Struct.* 2000;48:231–244.
9. Nguyen-Thoi T, Luong-Van H, Phung-Van P, Rabczuk T, Tran-Trung D. Dynamic responses of composite plates on the Pasternak foundation subjected to a moving mass by a cell-based smoothed discrete shear gap (CS-FEM-DSG3) method. *Int. J. Compos. Mat.* 2013;3:19–27.
10. Nguyen-Thoi T, Rabczuk T, Lam-Phat T, Ho-Huu V, Phung-Van P. Free vibration analysis of cracked Mindlin plate using an extended cell-based smoothed discrete shear gap method (XCS-DSG3). *Theoret. Appl. Fract. Mech.* 2014;72:150–163.
11. Nguyen-Thoi T, Bui-Xuan T, Phung-Van P, Nguyen-Hoang S, Nguyen-Xuan H. An edge-based smoothed three-node Mindlin plate element (ES-MIN3) for static and free vibration analyses of plates. *KSCE J. Civil Eng.* 2014;18:1072–1082.
12. Nguyen XH, Nguyen NT, Abdel Wahab M, Bordas SPA, Nguyen-Xuan H, Voa PT. A refined quasi-3D isogeometric analysis for functionally graded microplates based on the modified couple stress theory. *Comput. Meth. Appl. Mech. Eng.* 2017;313:904–940.
13. Noor AK. Stability of multilayered composite plates. *Fibre Sci. Tech.* 1975;8:81–89.
14. Phung-Van P, Luong-Van H, Nguyen-Thoi T, Nguyen-Xuan H. A cell-based smoothed discrete shear gap method (CS-FEM-DSG3) based on the C0-type higher-order shear deformation theory for dynamic responses of Mindlin plates on viscoelastic foundations subjected to a moving sprung vehicle. *Int. J. Numeric. Meth. Eng.* 2014;98:988–1014.
15. Phung-Van P, Nguyen-Thoi T, Luong-Van H, Thai-Hoang C, Nguyen-Xuan H. A cell-based smoothed discrete shear gap method (CS-FEM-DSG3) using layerwise deformation theory for dynamic response of composite plates resting on viscoelastic foundation. *Comput. Meth. Appl. Mech. Eng.* 2014;272:138–159.
16. Phung-Van P, Thai CH, Nguyen-Thoi T, Nguyen-Xuan H. Static and free vibration analyses of composite and sandwich plates by an edge-based smoothed discrete shear gap method (ES-DSG3) using triangular elements based on layerwise theory. *Compos. Part B: Eng.* 2014;60:227–238.
17. Phung Van P, Abdel Wahab M, Liew KM, Bordas SPA, Nguyen-Xuan H. Isogeometric analysis of functionally graded carbon nanotube-reinforced composite plates using higher-order shear deformation theory. *Compos. Struct.* 2015;123:137–149.

18. Phung Van P, Nguyen LB, Tran VL, Dinh TD, Thai CH, Bordas SPA, Abdel-Wahab M, Nguyen-Xuan H. An efficient computational approach for control of nonlinear transient responses of smart piezoelectric composite plates. *Int. J. Non-Linear Mech.* 2015;76:190–202.

19. Phung-Van P, De Lorenzis L, Thai CH, Abdel-Wahab M, Nguyen-Xuan H. Analysis of laminated composite plates integrated with piezoelectric sensors and actuators using higher-order shear deformation theory and isogeometric finite elements. *Comput. Mat. Sci.* 2015;96:495–505.

20. Phung-Van P, Tran LV, Ferreira AJM, Nguyen-Xuan H, Abdel-Wahab M. Nonlinear transient isogeometric analysis of smart piezoelectric functionally graded material plates based on generalized shear deformation theory under thermo-electro-mechanical loads. *Nonlinear Dyn.* 2016. https://doi.org/10.1007/s11071-016-3085-6.

21. Putcha NS, Reddy JN. Stability and natural vibration analysis of laminated plates by using a mixed element based on a refined plate theory. *J. Sound Vib.* 1986;104:285–300.

22. Reddy JN. A simple higher order theory for laminated composite plates. *J. Appl. Mech.* 1984;51:745–752.

23. Saidi H, Tounsi A, Bousahla AA. A simple hyperbolic shear deformation theory for vibration analysis of thick functionally graded rectangular plates resting on elastic foundations. *Geomech. Eng.* 2016;11:289–307.

24. Sheng GG, Wang X. Response and control of functionally graded laminated piezoelectric shells under thermal shock and moving loadings. *Compos. Struct.* 2010;93:132–141.

25. Shi DL, Feng XQ. The effect of nanotube waviness and agglomeration on the elastic property of carbon nanotube-reinforced composites. *J. Eng. Mat. Tech. ASME.* 2004;126:250–270.

26. Tran VL, Lee J, Ly HA, Abdel Wahab M, Nguyen-Xuan H. Vibration analysis of cracked FGM plates using higher-order shear deformation theory and extended isogeometric approach. *Int. J. Mech. Sci.* 2015;96–97:65–78.

27. Tran VL, Lee J, Nguyen-Van H, Nguyen-Xuan H, Abdel Wahab M. Geometrically nonlinear isogeometric analysis of laminated composite plates based on higher-order shear deformation theory. *Int. J. Non-Linear Mech.* 2015;72:42–52.

28. Tran LV, Phung-Van P, Lee J, Abdel Wahab A, Nguyen-Xuan H. Isogeometric analysis for nonlinear thermomechanical stability of functionally graded plates. *Compos. Struct.* 2016;140:655–667.

29. Thai CH, Ferreira AJM, Abdel Wahab M, Nguyen-Xuan H. A generalized layerwise higher-order shear deformation theory for laminated composite and sandwich plates based on isogeometric analysis. *Acta Mech.* 2016;227:1225–1250.

30. Thai C, Zenkour AM, Abdel Wahab M, Nguyen-Xuan H. A simple four-unknown shear and normal deformations theory for functionally graded isotropic and sandwich plates based on isogeometric analysis. *Compos. Struct.* 2016;139:77–95.

31. Zhu J, Li LY. A stiffened plate buckling model for calculating critical stress of distortional buckling of CFS beams. *Int. J. Mech. Sci.* 2016;115–116:457–464.

7 Vibration of Nanocomposite Plates

7.1 INTRODUCTION

The application and use of laminated composite structures in different industries such as aerospace, automobile, etc. is growing. This is due to the fact that in the laminated composite structure, the strength with respect to the weight is high, and these materials can improve the stability of the structure and decrease the weight of the system. So, in recent years, the study and mechanical analysis of laminated composite structures have been intense interests for researchers. In this chapter, the laminas are reinforced with CNTs, which can increase the stiffness of the structure.

Composite structures find extensive applications in various engineering fields due to their high strength-to-weight ratio and tailorable mechanical properties. Understanding the behavior of composite plates, especially under different loading and environmental conditions, is crucial for their optimal design and performance. Over the years, researchers have made significant contributions to the analysis of composite plates, focusing on aspects such as buckling, vibration, and dynamic stability. Afsharmanesh et al. [1] investigated the buckling and vibration characteristics of laminated composite circular plates on a Winkler-type foundation, shedding light on their structural responses under different loading conditions. Similarly, Akhavan et al. [2] provided exact solutions for rectangular Mindlin plates under in-plane loads resting on a Pasternak elastic foundation, enhancing the understanding of buckling phenomena in such structures. Dynamic stability analysis of laminated composite plates in thermal environments was explored by Chen et al. [3], offering insights into the structural response of composite plates subjected to temperature variations. Dash and Singh [4] studied the buckling and postbuckling behavior of laminated composite plates, contributing valuable knowledge to the understanding of their stability under mechanical loads. The nonlinear dynamic buckling of laminated angle-ply composite spherical caps was investigated by Gupta et al. [5], highlighting the complex behavior of composite structures under dynamic loading conditions. Javed et al. [6] focused on the vibration analysis of anti-symmetric angle-ply laminated plates under higher-order shear theory, providing a deeper understanding of their dynamic response. Recent research has also delved into advanced composite materials and structures. Kiani [7] examined the buckling behavior of functionally graded carbon nanotube (FG-CNT)-reinforced composite plates subjected to parabolic loading, offering insights into the mechanical responses of such advanced materials. Kolahchi et al. [8] proposed a nonlocal nonlinear analysis for buckling in embedded FG-SWCNT-reinforced micro plates under a magnetic field, contributing to the understanding of nanoscale composite structures. The dynamic stability analysis of temperature-dependent, functionally graded, CNT-reinforced visco-plates resting on

DOI: 10.1201/9781003510710-7

an orthotropic elastomeric medium was investigated by Kolahchi et al. [9], providing valuable information on the dynamic behavior of functionally graded composite structures under thermal effects. Furthermore, Kolahchi et al. [10] developed differential cubature and quadrature–Bolotin methods for the dynamic stability analysis of embedded piezoelectric nanoplates based on visco-nonlocal-piezoelasticity theories, offering new perspectives on the stability of smart composite structures. Continuing in this line of research, Kolahchi et al. [11] proposed visco-nonlocal-refined zigzag theories for the dynamic buckling of laminated nanoplates, using differential cubature–Bolotin methods, further advancing the understanding of dynamic instability in nanoscale composite structures. Moreover, the inclusion of delaminations in laminated composite plates was investigated by Lee and Park [12], providing insights into the effect of delamination on the buckling behavior of composite structures. Lei et al. [13, 14] explored the vibration and buckling characteristics of FG-CNT-reinforced thick composite quadrilateral and skew plates resting on Pasternak foundations, employing element-free approaches to enhance the analysis of complex composite structures. Madani et al. [15] employed the differential cubature method for the vibration analysis of embedded FG-CNT-reinforced piezoelectric cylindrical shells, providing valuable insights into the dynamic behavior of multifunctional composite structures. Mayandi and Jeyaraj [16] investigated the bending, buckling, and free vibration characteristics of FG-CNT-reinforced polymer composite beams under non-uniform thermal load, contributing to the understanding of the thermal response of advanced composite structures. Additionally, Norouzi and Younesian [17] studied the nonlinear vibration of laminated composite plates subjected to subsonic flow and external loads, offering insights into the dynamic behavior of composite structures under aerodynamic loads. Reddy [18] proposed a simple higher-order theory for laminated composite plates, offering a concise yet effective approach to analyzing the structural behavior of composite plates under various loading conditions. Singh and Chakrabarti [19] presented an efficient C0 finite element model for the buckling analysis of laminated composite plates, contributing to the development of robust computational techniques for structural analysis. Singh et al. [20] employed meshless collocations for the buckling analysis of laminated composite plates subjected to mechanical and thermal loads, showcasing the versatility of numerical methods in structural analysis. Wang et al. [21] investigated hygrothermal effects on the dynamic instability of laminated plates under arbitrary pulsating loads, highlighting the importance of considering environmental factors in structural design. Yang et al. [22] studied the dynamic buckling of thermo-electro-mechanically loaded FG-CNTRC beams, providing valuable insights into the multifunctional behavior of advanced composite structures. Yas and Heshmati [23] analyzed the dynamic behavior of functionally graded nanocomposite beams reinforced by randomly oriented carbon nanotubes under the action of moving loads, contributing to the understanding of dynamic responses in nanoscale composite structures. Additionally, Zhang and Liew [24] conducted a large deflection analysis of FG-CNT-reinforced composite skew plates resting on Pasternak foundations using an element-free approach. This study contributes to the understanding of the structural behavior of composite plates under complex loading conditions, offering insights into the influence of geometric nonlinearity on the performance of such structures. These studies underscore the diverse approaches and methodologies employed in analyzing the buckling, vibration, and dynamic stability of composite plates, catering to the evolving needs of structural engineering and materials science. By combining theoretical developments with advanced computational techniques, researchers continue to expand the boundaries of knowledge in composite mechanics, laying the groundwork for innovative structural design and optimization strategies.

However, the buckling analysis of sandwich nanocomposite plates has not been studied by researchers. In this chapter, the buckling load of sandwich plates with FG-CNT-reinforced layers is obtained based on the Reddy shear deformation plate theory. The material properties of the plates and constants of the elastic medium are assumed temperature dependent. The mixture rule is utilized to calculate the equivalent characteristics of the nanocomposite structure. The effects of different parameters such as volume percent and distribution types of the CNTs, temperature change, elastic medium, magnetic field, and geometrical parameters of the plates on the buckling load of the sandwich structure and the buckling behavior of the structure are shown.

7.2 MOTION EQUATIONS

A sandwich plate with FG-CNT-reinforced layers resting on an elastic medium is shown in Figure 7.1 with length a, width b, and thickness h.

Based on the Reddy plate theory presented in Chapter 1, the displacement field can be assumed in this chapter. Based on this theory, the stress–strain relations can be written using Hook's law as follows:

$$\begin{Bmatrix} \sigma_{xx} \\ \sigma_{yy} \\ \sigma_{xy} \\ \sigma_{xz} \\ \sigma_{yz} \end{Bmatrix} = \begin{bmatrix} Q_{11} & Q_{12} & 0 & 0 & 0 \\ Q_{12} & Q_{22} & 0 & 0 & 0 \\ 0 & 0 & Q_{66} & 0 & 0 \\ 0 & 0 & 0 & Q_{55} & 0 \\ 0 & 0 & 0 & 0 & Q_{44} \end{bmatrix} \begin{Bmatrix} \varepsilon_{xx} - \alpha_{xx}T \\ \varepsilon_{yy} - \alpha_{yy}T \\ \gamma_{xy} \\ \gamma_{xz} \\ \gamma_{yz} \end{Bmatrix}, \tag{7.1}$$

Where $(\alpha_{xx}, \alpha_{yy})$ and T are thermal expansion and temperature change, respectively; Q_{ij} $(i, j = 1, 2, ..., 6)$ denotes elastic coefficients, which can be obtained by the mixture rule.

According to this theory, the effective Young's and shear moduli of the structure may be expressed as [24]:

$$E_{11} = \eta_1 V_{CNT} E_{r11} + (1 - V_{CNT}) E_m, \tag{7.2}$$

$$\frac{\eta_2}{E_{22}} = \frac{V_{CNT}}{E_{r22}} + \frac{(1 - V_{CNT})}{E_m}, \tag{7.3}$$

$$\frac{\eta_3}{G_{12}} = \frac{V_{CNT}}{G_{r12}} + \frac{(1 - V_{CNT})}{G_m}, \tag{7.4}$$

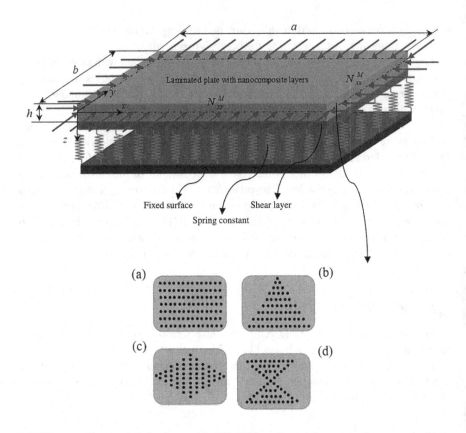

FIGURE 7.1 The sandwich plate with FG-CNT-reinforced layers resting on an elastic foundation.

where E_{r11}, E_{r22}, and E_m are Young's moduli of the CNTs and matrix, respectively; G_{r11} and G_m are the shear moduli of the CNTs and matrix, respectively; V_{CNT} and V_m show the volume fractions of the CNTs and matrix, respectively; and η_j ($j = 1, 2, 3$) is the CNT efficiency parameter for considering the size-dependent material properties. Note that this parameter may be calculated using molecular dynamics (MD). However, the CNT distribution for the mentioned patters obeys from the following relations:

$$UD: \quad V_{CNT} = V_{CNT}^*, \tag{7.5}$$

$$FGV: \quad V_{CNT}(z) = \left(1 + \frac{2z}{h}\right)V_{CNT}^*, \tag{7.6}$$

$$FGO: \quad V_{CNT}(z) = 2\left(1 - \frac{2|z|}{h}\right)V_{CNT}^*, \tag{7.7}$$

$$FGX: \quad V_{CNT}(z) = 2\left(\frac{2|z|}{h}\right)V_{CNT}^*. \tag{7.8}$$

Furthermore, the thermal expansion coefficients in the axial and transverse directions (α_{11} and α_{22}, respectively) and the density (ρ) of the nanocomposite structure can be written as:

$$\rho = V_{CNT}\rho_r + V_{m}\rho_{m}, \tag{7.9}$$

$$\alpha_{11} = V_{CNT}\alpha_{r11} + V_{m}\alpha_{m}, \tag{7.10}$$

$$\alpha_{22} = \left(1+\nu_{r12}\right)V_{CNT}\alpha_{r22} + \left(1+\nu_{m}\right)V_{m}\alpha_{m} - \nu_{12}\alpha_{11}, \tag{7.11}$$

where

$$V_{CNT}^* = \frac{w_{CNT}}{w_{CNT} + \left(\rho_{CNT}/\rho_{m}\right) - \left(\rho_{CNT}/\rho_{m}\right)w_{CNT}}, \tag{7.12}$$

where w_{CNT} is the mass fraction of the CNTs; ρ_{m} and ρ_{CNT} represent the densities of the matrix and CNTs, respectively; ν_{r12} and ν_{m} are the Poisson's ratios of the CNT and matrix, respectively; and (α_{r11}, α_{r22}) and α_{m} are the thermal expansion coefficients of the CNT and matrix, respectively. Note that ν_{12} is assumed constant.

The strain energy of the structure can be written as:

$$U = \frac{1}{2}\int_{\Omega_0}\int_{-h/2}^{h/2}\left(\sigma_{xx}\varepsilon_{xx} + \sigma_{yy}\varepsilon_{yy} + \sigma_{xy}\gamma_{xy} + \sigma_{xz}\gamma_{xz} + \sigma_{yz}\gamma_{yz}\right)dV. \tag{7.13}$$

Substituting Eq. (7.1) into Eq. (7.13) yields:

$$U = \frac{1}{2}\int_0^b\int_0^a\left[N_{xx}\left(\frac{\partial u}{\partial x}\right) + N_{yy}\left(\frac{\partial v}{\partial y}\right) + Q_{yy}\left(\frac{\partial w}{\partial y} + \varphi_y\right) + Q_{xx}\left(\frac{\partial w}{\partial x} + \varphi_x\right) + N_{xy}\left(\frac{\partial v}{\partial x} + \frac{\partial u}{\partial y}\right)\right.$$

$$+ M_{xx}\frac{\partial\varphi_x}{\partial x} + M_{yy}\frac{\partial\varphi_y}{\partial y} + M_{xy}\left(\frac{\partial\varphi_x}{\partial y} + \frac{\partial\varphi_y}{\partial x}\right) + K_{yy}\left(c_2\left(\varphi_y + \frac{\partial w}{\partial y}\right)\right) + K_{xx}\left(c_2\left(\varphi_x + \frac{\partial w}{\partial x}\right)\right),$$

$$\left. + P_{xx}\left(c_1\left(\frac{\partial\varphi_x}{\partial x} + \frac{\partial^2 w}{\partial x^2}\right)\right) + P_{yy}\left(c_1\left(\frac{\partial\varphi_y}{\partial y} + \frac{\partial^2 w}{\partial y^2}\right)\right) + P_{xy}\left(\frac{\partial\varphi_y}{\partial x} + \frac{\partial\varphi_x}{\partial y} + 2\frac{\partial^2 w}{\partial x\partial y}\right)\right]dxdy,$$

$$\tag{7.14}$$

where the stress resultants can be defined as:

$$\left[\left(N_{xx},N_{yy},N_{xy}\right),\left(M_{xx},M_{yy},M_{xy}\right),\left(P_{xx},P_{yy},P_{xy}\right)\right] = \int_{-h/2}^{h/2}\begin{bmatrix}\sigma_{xx}\\\sigma_{yy}\\\sigma_{xy}\end{bmatrix}\left[1,z,z^3\right]dz. \tag{7.15}$$

The work due to the in-plane external loads, elastic medium, and magnetic field can be expressed as [11]:

$$W = -\frac{1}{2}\int_0^b\int_0^a\left[\left(N_{xx}^M + N_{xx}^T\right)\left(\frac{\partial w}{\partial x}\right)^2 + \left(N_{yy}^M + N_{yy}^T\right)\left(\frac{\partial w}{\partial y}\right)^2\right]dxdy - \frac{1}{2}\int_0^b\int_0^a\left[\eta hH_x^2\frac{\partial^2 w}{\partial x^2}\right]dxdy$$

$$-\frac{1}{2}\int_0^b\int_0^a\left[\begin{array}{l}-K_w w + G_\xi\left(\cos^2\theta\dfrac{\partial^2 w}{\partial x^2} + 2\cos\theta\sin\theta\dfrac{\partial^2 w}{R\partial x\partial\theta} + \sin^2\theta\dfrac{\partial^2 w}{R^2\partial\theta^2}\right)\\[3mm] +G_\eta\left(\sin^2\theta\dfrac{\partial^2 w}{\partial x^2} - 2\sin\theta\cos\theta\dfrac{\partial^2 w}{R\partial x\partial\theta} + \cos^2\theta\dfrac{\partial^2 w}{R^2\partial\theta^2}\right)\end{array}\right]dxdy, \qquad (7.17)$$

where K_w and (G_ξ, G_ι) are Winkler's spring modulus and shear layer coefficients, respectively. In addition, angle θ describes the local ξ direction of the orthotropic foundation with respect to the global x-axis of the plate; η is the magnetic permeability; and H_x is the magnetic field. Also, $N_{xx}^M = -p$ and $N_{yy}^M = \alpha N_{xx}^M$ are loads applied to the plate in x and y directions, respectively, and α is a constant coefficient. In addition, $N_{\theta\theta}^T, N_{xx}^T$ are thermal forces, which may be written as:

$$\begin{Bmatrix}N_{xx}^T\\N_{\theta\theta}^T\end{Bmatrix} = \int_{-h/2}^{h/2}\begin{Bmatrix}C_{11}(T,z)\alpha_{xx}(z) + C_{12}(T,z)\alpha_{\theta\theta}(z)\\C_{12}(T,z)\alpha_{xx}(z) + C_{22}(T,z)\alpha_{\theta\theta}(z)\end{Bmatrix}\Delta T\,dz. \qquad (7.18)$$

The governing equations can be derived by Hamilton's principal as follows:

$$\delta u: \frac{\partial N_{xx}}{\partial x} + \frac{\partial N_{xy}}{\partial y} = 0, \qquad (7.19)$$

$$\delta v: \frac{\partial N_{xy}}{\partial x} + \frac{\partial N_{yy}}{\partial y} = 0, \qquad (7.20)$$

$$\delta w: \frac{\partial Q_{xx}}{\partial x} + \frac{\partial Q_{yy}}{\partial y} + c_2\left(\frac{\partial K_{xx}}{\partial x} + \frac{\partial K_{yy}}{\partial y}\right) + \left(N_{xx}^M + N_{xx}^T\right)\frac{\partial^2 w}{\partial x^2} + \left(N_{yy}^M + N_{yy}^T\right)\frac{\partial^2 w}{\partial y^2}$$

$$-c_1\left(\frac{\partial^2 P_{xx}}{\partial x^2} + 2\frac{\partial^2 P_{xy}}{\partial x\partial y} + \frac{\partial^2 P_{yy}}{\partial y^2}\right) - K_w w + G_\xi\left(\cos^2\theta\frac{\partial^2 w}{\partial x^2} + 2\cos\theta\sin\theta\frac{\partial^2 w}{R\partial x\partial\theta} + \sin^2\theta\frac{\partial^2 w}{R^2\partial\theta^2}\right)$$

$$+G_\eta\left(\sin^2\theta\frac{\partial^2 w}{\partial x^2} - 2\sin\theta\cos\theta\frac{\partial^2 w}{R\partial x\partial\theta} + \cos^2\theta\frac{\partial^2 w}{R^2\partial\theta^2}\right) + \eta hH_x^2\frac{\partial^2 w}{\partial x^2} = 0, \qquad (7.21)$$

$$\delta\phi_x: \frac{\partial M_{xx}}{\partial x} + \frac{\partial M_{xy}}{\partial y} + c_1\left(\frac{\partial P_{xx}}{\partial x} + \frac{\partial P_{xy}}{\partial y}\right) - Q_{xx} - c_2 K_{xx} = 0, \qquad (7.22)$$

$$\delta\phi_y: \frac{\partial M_{xy}}{\partial x} + \frac{\partial M_{yy}}{\partial y} + c_1\left(\frac{\partial P_{xy}}{\partial x} + \frac{\partial P_{yy}}{\partial y}\right) - Q_{yy} - c_2 K_{yy} = 0. \qquad (7.23)$$

Combining Eqs. (7.1), (7.15), and (7.16), the stress resultants can be obtained as follows:

$$N_{xx} = A_{11}\frac{\partial u}{\partial x} + A_{12}\frac{\partial v}{\partial y} + B_{11}\frac{\partial \varphi_x}{\partial x} + B_{12}\frac{\partial \varphi_y}{\partial y} + E_{11}c_1\left(\frac{\partial \varphi_x}{\partial x} + \frac{\partial^2 w}{\partial x^2}\right) + E_{12}c_1\left(\frac{\partial \varphi_y}{\partial y} + \frac{\partial^2 w}{\partial y^2}\right),$$

$$(7.24)$$

$$N_{yy} = A_{12}\frac{\partial u}{\partial x} + A_{22}\frac{\partial v}{\partial y} B_{12}\frac{\partial \varphi_x}{\partial x} + B_{22}\frac{\partial \varphi_y}{\partial y} + E_{12}c_1\left(\frac{\partial \varphi_x}{\partial x} + \frac{\partial^2 w}{\partial x^2}\right) + E_{22}c_1\left(\frac{\partial \varphi_y}{\partial y} + \frac{\partial^2 w}{\partial y^2}\right),$$

$$(7.25)$$

TABLE 7.1

Material Properties of CNTs with a Length of 9.26 nm and Thickness of $h = 0.067$ nm

Temperature (K)	E_{11}^{CNT} TPa	E_{22}^{CNT} TPa	G_{12}^{CNT} TPa	α_{12}^{CNT} $(10^{-6}/K)$	α_{22}^{CNT} $(10^{-6}/K)$
300	5.6466	7.0800	1.9445	3.4584	5.1682
500	5.5308	6.9348	1.9643	4.5361	5.0189
700	5.4744	6.8641	1.9644	4.6677	4.8943

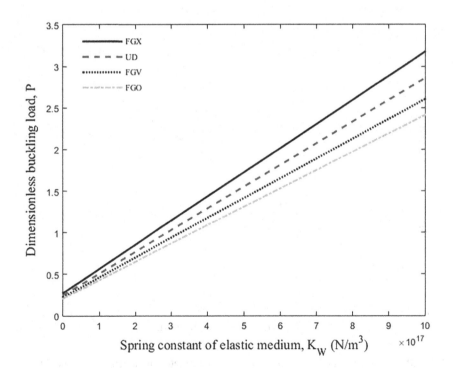

FIGURE 7.2 Effects of CNT distribution type on the variation in dimensionless buckling load versus spring constant of an elastic medium.

$$N_{xy} = A_{66}\left(\frac{\partial u}{\partial y} + \frac{\partial v}{\partial x}\right) + B_{66}\left(\frac{\partial \varphi_x}{\partial y} + \frac{\partial \varphi_y}{\partial x}\right) + E_{66}c_1\left(\frac{\partial \varphi_y}{\partial x} + \frac{\partial \varphi_x}{\partial y} + 2\frac{\partial^2 w}{\partial x \partial y}\right), \quad (7.26)$$

$$M_{xx} = B_{11}\frac{\partial u}{\partial x} + B_{12}\frac{\partial v}{\partial y} + D_{11}\frac{\partial \varphi_x}{\partial x} + D_{12}\frac{\partial \varphi_y}{\partial y} + F_{11}c_1\left(\frac{\partial \varphi_x}{\partial x} + \frac{\partial^2 w}{\partial x^2}\right) + F_{12}c_1\left(\frac{\partial \varphi_y}{\partial y} + \frac{\partial^2 w}{\partial y^2}\right),$$

$$(7.27)$$

$$M_{yy} = B_{12}\frac{\partial u}{\partial x} + B_{22}\frac{\partial v}{\partial y} + D_{12}\frac{\partial \varphi_x}{\partial x} + D_{22}\frac{\partial \varphi_y}{\partial y} + F_{12}c_1\left(\frac{\partial \varphi_x}{\partial x} + \frac{\partial^2 w}{\partial x^2}\right) + F_{22}c_1\left(\frac{\partial \varphi_y}{\partial y} + \frac{\partial^2 w}{\partial y^2}\right),$$

$$(7.28)$$

$$M_{xy} = B_{66}\left(\frac{\partial u}{\partial y} + \frac{\partial v}{\partial x}\right) + D_{66}\left(\frac{\partial \varphi_x}{\partial y} + \frac{\partial \varphi_y}{\partial x}\right) + F_{66}c_1\left(\frac{\partial \varphi_y}{\partial x} + \frac{\partial \varphi_x}{\partial y} + 2\frac{\partial^2 w}{\partial x \partial y}\right), \quad (7.29)$$

$$P_{xx} = E_{11}\frac{\partial u}{\partial x} + E_{12}\frac{\partial v}{\partial y} + F_{11}\frac{\partial \varphi_x}{\partial x} + F_{12}\frac{\partial \varphi_y}{\partial y} + H_{11}c_1\left(\frac{\partial \varphi_x}{\partial x} + \frac{\partial^2 w}{\partial x^2}\right) + H_{12}c_1\left(\frac{\partial \varphi_y}{\partial y} + \frac{\partial^2 w}{\partial y^2}\right),$$

$$(7.30)$$

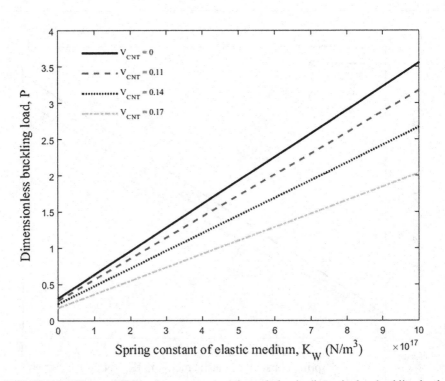

FIGURE 7.3 Effects of CNT volume percent on the variation in dimensionless buckling load versus spring constant of an elastic medium.

$$P_{yy} = E_{12}\frac{\partial u}{\partial x} + E_{22}\frac{\partial v}{\partial y} + F_{12}\frac{\partial \varphi_x}{\partial x} + F_{22}\frac{\partial \varphi_y}{\partial y} + H_{12}c_1\left(\frac{\partial \varphi_x}{\partial x} + \frac{\partial^2 w}{\partial x^2}\right) + H_{22}c_1\left(\frac{\partial \varphi_y}{\partial y} + \frac{\partial^2 w}{\partial y^2}\right),$$

(7.31)

$$P_{xy} = E_{66}\left(\frac{\partial u}{\partial y} + \frac{\partial v}{\partial x}\right) + F_{66}\left(\frac{\partial \varphi_x}{\partial y} + \frac{\partial \varphi_y}{\partial x}\right) + H_{66}c_1\left(\frac{\partial \varphi_y}{\partial x} + \frac{\partial \varphi_x}{\partial y} + 2\frac{\partial^2 w}{\partial x \partial y}\right),$$ (7.32)

$$Q_{xx} = A_{55}\left(\frac{\partial w}{\partial x} + \varphi_x\right) + D_{55}c_2\left(\varphi_x + \frac{\partial w}{\partial x}\right),$$

(7.33)

$$Q_{yy} = A_{44}\left(\frac{\partial w}{\partial y} + \varphi_y\right) + D_{44}c_2\left(\frac{\partial w}{\partial y} + \varphi_y\right),$$

(7.34)

$$K_{xx} = D_{55}\left(\frac{\partial w}{\partial x} + \varphi_x\right) + F_{55}c_2\left(\varphi_x + \frac{\partial w}{\partial x}\right),$$

(7.35)

$$K_{yy} = D_{45}\left(\frac{\partial w}{\partial y} + \varphi_y\right) + F_{44}c_2\left(\frac{\partial w}{\partial y} + \varphi_y\right),$$

(7.36)

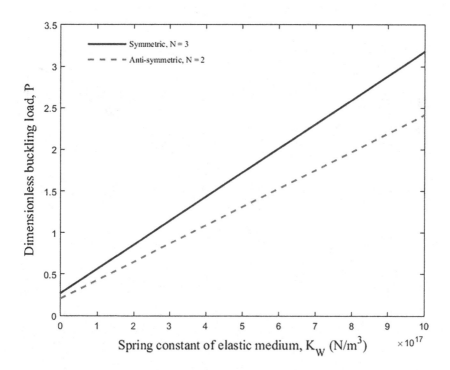

FIGURE 7.4 Effects of number of laminas on the variation in dimensionless buckling load versus spring constant of an elastic medium.

where

$$\left(A_{ij}, B_{ij}, D_{ij}, B_{ij}, F_{ij}, H_{ij}\right) = \sum_{k=1}^{N} \int_{z^{(k-1)}}^{z^{(k)}} Q_{ij}^{(k)}\left(1, z, z^2, z^3, z^4, z^6\right) dz, \qquad (i, j = 1, 2, 6), \quad (7.37)$$

in which N is the number of sandwich plate layers. Finally, the governing equations are obtained by substituting Eqs. (7.24)–(7.36) into the governing equations. Using the Navier method presented in Chapter 2, we have:

$$[K] + p[K_G] = [0], \tag{7.38}$$

where $[K]$ and $[K_G]$ are stiffness and geometric matrixes, respectively. Finally, the buckling load of the system (P) can be calculated by using an eigenvalue problem, which is discussed in the next section.

7.3 NUMERICAL RESULTS

In this section, the numerical results of the buckling analysis of FG-CNT-reinforced sandwich plates resting on an orthotropic temperature-dependent elastic foundation are

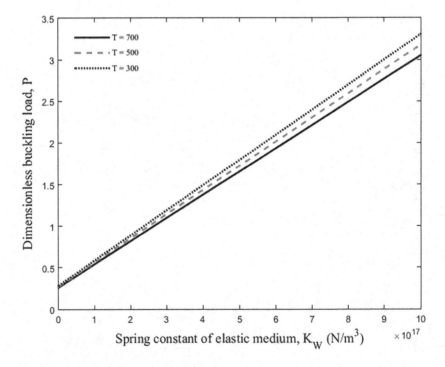

FIGURE 7.5 Effects of temperature change on the variation in dimensionless buckling load versus spring constant of the elastic medium.

presented. Each layer of the sandwich structure is made from poly methyl methacrylate (PMMA) with the constant Poisson's ratio of $\nu_m = 0.34$, temperature-dependent thermal coefficient of $\alpha_m = (1+0.0005\Delta T)\times 10^{-6}/K$, and temperature-dependent Young's modulus of $E_m = (3.52 - 0.0034T)$ GPa in which $T = T_0 + \Delta T$ and $T_0 = 300$ K (room temperature). CNTs as reinforcement of the lamina layers have the material properties listed in Table 7.1 [24]. Since the surrounding medium is relatively soft, the foundation stiffness K_w may be expressed by [24]:

$$K_w = \frac{E_0}{4L(1-\nu_0^2)(2-c_1)^2}\Big[5-\big(2\gamma_1^2 +6\gamma_1 +5\big)\exp(-2\gamma_1)\Big],\qquad(7.39)$$

where

$$c_1 = (\gamma_1 + 2)\exp(-\gamma_1),\qquad(7.40)$$

$$\gamma_1 = \frac{H_s}{L},\qquad(7.41)$$

$$E_0 = \frac{E_s}{\left(1-\nu_s^2\right)},\qquad(7.42)$$

$$\nu_0 = \frac{\nu_s}{\left(1-\nu_s\right)},\qquad(7.43)$$

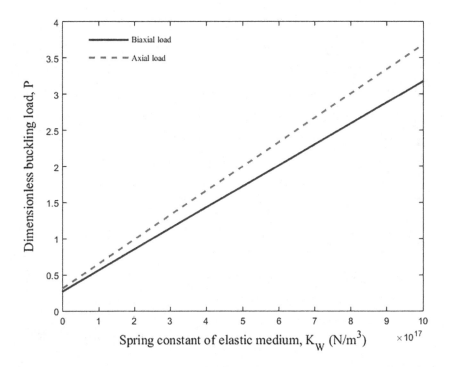

FIGURE 7.6 Effects of loading type on the variation in dimensionless buckling load versus spring constant of the elastic medium.

where E_s, ν_s, and H_s are the Young's modulus, Poisson's ratio, and depth of the foundation, respectively. In this chapter, E_s is assumed to be temperature dependent, while ν_s is assumed to be a constant. The elastic medium is made of poly dimethylsiloxane (PDMS), the temperature-dependent material properties of which are assumed to be $\nu_s = 0.48$ and $E_s = (3.22 - 0.0034T)\,GPa$, in which $T = T_0 + \Delta T$ and $T_0 = 300\ K$ (room temperature) [9].

The effect of the distribution type of CNTs in layers of sandwich plates on the dimensionless buckling load ($P = p\,/\,E_{11}^m h$) of the system versus the spring constant of the elastic medium is presented in Figure 7.2. The CNT uniform distribution and three types of FG patterns, namely, FG-V type, FG-O type, and FG-X type, are considered. It can be seen that the buckling load increases with an increasing spring constant of the elastic medium. This is because increasing the spring constant of the elastic medium leads to a stiffer structure. With respect to the distribution types of CNTs in the sandwich plate, it can be concluded that the FGX pattern is the best choice compared to other cases. This is because, in the FGX mode, the buckling load is maximum, which means the stiffness of the system is higher with respect to the other three patterns. Meanwhile, the buckling load of the structure with uniform CNT distribution is higher than that in the FGV and FGO models. However, it can be concluded that the CNTs distributed close to the top and bottom are more efficient than those distributed near the mid-plane.

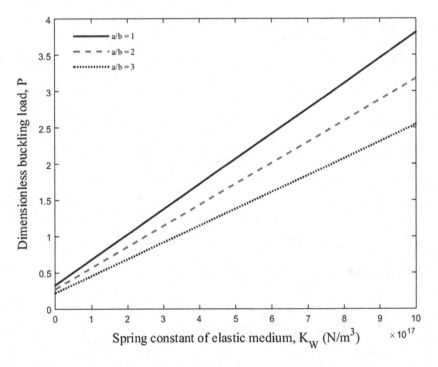

FIGURE 7.7 Effects of length-to-width ratio on the variation in dimensionless buckling load versus spring constant of the elastic medium.

The effect of the CNT volume fraction on the dimensionless buckling load of the sandwich plate with respect to the spring constant of the elastic medium is shown in Figure 7.3. It can be seen that increasing the CNT volume fraction increases the dimensionless buckling load of the structure. This is due to the fact that the increase in CNT volume fraction leads to a harder structure. It is also concluded that the effects of CNT volume fraction become more prominent at a higher spring constant of the elastic foundation.

Figure 7.4 illustrates the variation in the dimensionless buckling load versus the spring constant of an elastic medium for the symmetric and anti-symmetric sandwich plates. It can be seen that in symmetric laminated composites (with three layers), the dimensionless buckling load increases compared with that for the anti-symmetric ones (with two layers). The reason is that the symmetric laminated composite plates are more balanced and stable.

The effect of temperature change on the dimensionless buckling load of the nanocomposite sandwich plate with respect to the spring constant of the elastic medium is demonstrated in Figure 7.5. As in the other figures, increasing the spring constant of the elastic medium increases the dimensionless buckling load of the structure. It can be also found that the dimensionless buckling load of the structure decreases with increasing temperature change, which is due to the higher stiffness of the nanocomposite sandwich plate with lower temperature.

Figure 7.6 examines the influence of the loading type on the dimensionless buckling load of the structure versus the spring constant of the elastic medium. Two types

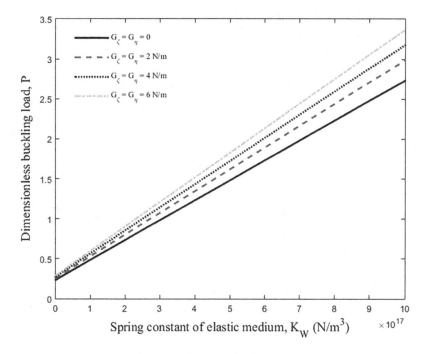

FIGURE 7.8 Effects of shear constant of the elastic medium on the variation in dimensionless buckling load versus spring constant of the elastic medium.

of loading, including axial (along the x-axis) and biaxial (along the x- and y-axes), are considered. As can be seen, for the biaxial loading type, the dimensionless buckling load is lower than that for the axial loading type. The reason is that, in the biaxial loading type, the load applied to the edges is higher than that for the axial loading type, and therefore, the buckling of the structure occurs sooner. Also the effect of loading type is apparent in the higher spring constant of the elastic medium.

The effect of the length-to-width ratio (a/b) on the dimensionless buckling load with respect to the spring constant of the elastic medium is depicted in Figure 7.7. As can be seen, the dimensionless buckling load of the sandwich plate decreases with an increasing length-to-width ratio. This is because increasing the length-to-width ratio leads to a softer structure. Meanwhile, the effect of length-to-width ratio on the dimensionless buckling load becomes more prominent at a higher spring constant of the elastic medium.

The effect of the shear constant of the elastic medium on the dimensionless buckling load versus spring constant of the elastic medium is shown in Figure 7.8. It can be observed that with increasing shear and spring constants of the elastic medium, the dimensionless buckling load is enhanced. This is physically due to the fact that when increasing the shear and spring constants of the elastic medium, the stiffness of the structure increases.

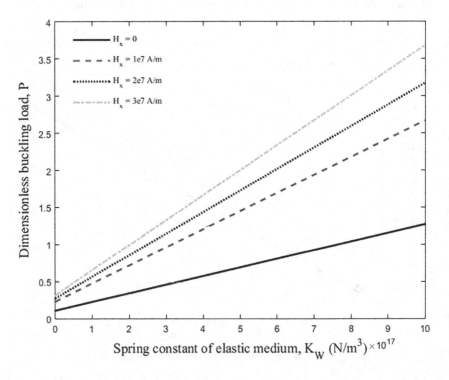

FIGURE 7.9 Effects of axial magnetic field on the variation in dimensionless buckling load versus spring constant of the elastic medium.

Figure 7.9 demonstrates the effect of axial magnetic field on the variation in dimensionless buckling load versus spring constant of the elastic medium. As can be seen, with an increasing axial magnetic field, the structure becomes stiffer, and consequently, the dimensionless buckling load is enhanced.

REFERENCES

1. Afsharmanesh B, Ghaheri A, Taheri-Behrooz F. Buckling and vibration of laminated composite circular plate on Winkler-type foundation. *Steel Compos. Struct.* 2014;17:1–19.
2. Akhavan H, Hosseini Hashemi Sh, Rokni Damavandi Taher H, Alibeigloo A, Vahabi Sh. Exact solutions for rectangular Mindlin plates under in-plane loads resting on Pasternak elastic foundation. Part I: Buckling analysis. *Comput. Mat. Sci.* 2009;44:968–978.
3. Chen CS, Tsai TC, Chen WR, Wei CL. Dynamic stability analysis of laminated composite plates in thermal environments. *Steel Compos. Struct.* 2013;15:57–79.
4. Dash P, Singh BN. Buckling and post-buckling of laminated composite plates. *Mech. Res. Commun.* 2012;46:1–7.
5. Gupta SS, Patel BP, Ganapathi M. Nonlinear dynamic buckling of laminated angle-ply composite spherical caps. *Struct. Eng. Mech.* 2003;15:463–476.
6. Javed S, Viswanathan KK, Aziz ZA, Karthik K, Lee JH. Vibration of antisymmetric angle-ply laminated plates under higher order shear theory. *Steel Compos. Struct.* 2016;22:1281–1299.
7. Kiani Y. *Buckling of FG-CNT-reinforced composite plates subjected to parabolic loading*. Acta Mechanica, 2016 (in press).
8. Kolahchi R, Rabani Bidgoli M, Beygipoor Gh, Fakhar MH. A nonlocal nonlinear analysis for buckling in embedded FG-SWCNT-reinforced microplates subjected to magnetic field. *J. Mech. Sci. Tech.* 2015;29:3669–3677.
9. Kolahchi R, Safari M, Esmailpour M. Dynamic stability analysis of temperature-dependent functionally graded CNT-reinforced visco-plates resting on orthotropic elastomeric medium. *Compos. Struct.* 2016;150:255–265.
10. Kolahchi R, Hosseini H, Esmailpour M. Differential cubature and quadrature-Bolotin methods for dynamic stability of embedded piezoelectric nanoplates based on visco-nonlocal-piezoelasticity theories. *Compos. Struct.* 2016;157:174–186.
11. Kolahchi R, Zarei MS, Hajmohammad MH, Naddaf Oskouei A. Visco-nonlocal-refined zigzag theories for dynamic buckling of laminated nanoplates using differential cubature-Bolotin methods. *Thin-Wall Struct.* 2017;113:162–169.
12. Lee SY, Park DY. Buckling analysis of laminated composite plates containing delaminations using the enhanced assumed strain solid element. *Int. J. Solids Struct.* 2007;44:8006–8027.
13. Lei ZX, Zhang LW, Liew KM. Vibration of FG-CNT reinforced composite thick quadrilateral plates resting on Pasternak foundations. *Eng. Anal. Bound. Elem.* 2016;64:1–11.
14. Lei ZX, Zhang LW, Liew KM. Buckling of FG-CNT reinforced composite thick skew plates resting on Pasternak foundations based on an element-free approach. *Appl. Math. Comput.* 2016;266:773–791.
15. Madani H, Hosseini H, Shokravi M. Differential cubature method for vibration analysis of embedded FG-CNT-reinforced piezoelectric cylindrical shells subjected to uniform and non-uniform temperature distributions. *Steel Compos. Struct.* 2016;22:889–913.
16. Mayandi K, Jeyaraj P. Bending, buckling and free vibration characteristics of FG-CNT-reinforced polymer composite beam under non-uniform thermal load. *Proceed. Inst. Mech. Eng. Part L.* 2015;229:13–28.
17. Norouzi H, Younesian D. Nonlinear vibration of laminated composite plates subjected to subsonic flow and external loads. *Steel Compos. Struct.* 2016;22:1261–1280.

18. Reddy JN. A simple higher order theory for laminated composite plates. *J. Appl. Mech.* 1984;51:745–752.
19. Singh SK, Chakrabarti A. Buckling analysis of laminated composite plates using an efficient C0 FE model. *Lat. Am. J. Solids Struct.* 2012;9:1–15.
20. Singh S, Singh J, Shukla KK. Buckling of laminated composite plates subjected to mechanical and thermal loads using meshless collocations. *J. Mech. Sci. Tech.* 2013;27:327–336.
21. Wang H, Chen CS, Fung CP. Hygrothermal effects on dynamic instability of a laminated plate under an arbitrary pulsating load. *Struct. Eng. Mech.* 2013;48:34–46.
22. Yang J, Ke LL, Feng C. Dynamic buckling of thermo-electro-mechanically loaded FG-CNTRC beams. *Int. J. Str. Stab. Dyn.* 2015;15:1540017.
23. Yas MH, Heshmati M. Dynamic analysis of functionally graded nanocomposite beams reinforced by randomly oriented carbon nanotube under the action of moving load. *Appl. Math. Model.* 2012;36:1371–1394.
24. Zhang LW, Liew KM. Large deflection analysis of FG-CNT reinforced composite skew plates resting on Pasternak foundations using an element-free approach. *Compos. Struct.* 2015;132:974–983.

8 Dynamic Buckling of Nanocomposite Plates

8.1 INTRODUCTION

Nanocomposite materials are made up of different functional components, such as polymers, nanoparticles, and ligands, with at least one component measuring in nanometers. A common challenge in these materials is achieving a uniform dispersion of nanoparticles within the polymer matrix. Often, nanoparticles tend to cluster together or separate from the polymer, leading to issues in processing the films and resulting in a high density of defects. Moreover, physical properties of the composite material are very sensitive to particle dispersion within the nanocomposite.

Abdollahzadeh Shahrbabaki and Alibeigloo [1] delved into the three-dimensional free vibration characteristics of carbon nanotube (CNT)-reinforced composite plates, exploring various boundary conditions using the Ritz method. Atteshamuddin and Yuwaraj [2] conducted a study on the free vibration behavior of angle-ply laminated composite plates and soft core sandwich plates, shedding light on their dynamic responses. Chetan et al. [3] developed a comprehensive model to simulate the interfacial damping effects caused by nanotube agglomerations in nanocomposite materials, a crucial aspect in understanding their mechanical properties. Eringen [4] contributed significantly by discussing the differential equations governing nonlocal elasticity and provided solutions for phenomena like screw dislocations and surface waves, laying a theoretical foundation for subsequent research in the field. Ke et al. [5] investigated the intricate behavior of nonlocal piezoelectric nanoplates under diverse boundary conditions, essential for designing advanced nanoelectromechanical systems. Kiani [6] explored the free vibration characteristics of conducting nanoplates subjected to in-plane magnetic fields using nonlocal shear deformable plate theories, offering insights into the dynamic behavior of nanocomposites under external stimuli. Lanhe et al. [7] conducted dynamic stability analysis on functionally graded material (FGM) plates using the moving least squares differential quadrature method, contributing to the understanding of structural stability in advanced composite materials. Lei et al. [8] studied the vibration characteristics of nonlocal Kelvin–Voigt viscoelastic damped Timoshenko beams, providing valuable information for applications in damping systems and vibration control. Li et al. [9] investigated the buckling and free vibration of magnetoelectroelastic nanoplates based on nonlocal theory, addressing the unique behavior of these materials under electromechanical coupling. Mantari and Guedes Soares [10] proposed a novel four-unknown quasi-3D shear deformation theory for advanced composite plates, offering a refined approach for accurately predicting the mechanical response of complex structures. Mori and Tanaka [11] discussed the average stress in the matrix and the elastic energy of materials with misfitting inclusions, enhancing the

DOI: 10.1201/9781003510710-8

understanding of composite materials with heterogeneous inclusions. Natarajan et al. [12] applied higher-order structural theory to analyze the bending and free vibration of sandwich plates reinforced with carbon nanotubes, providing insights into the structural performance of these advanced composite materials. Pandit et al. [13] employed an improved higher-order zigzag theory to investigate the behavior of laminated sandwich plates, offering a robust analytical framework for predicting their mechanical responses. Phung-Van et al. [14] conducted an isogeometric analysis of functionally graded carbon nanotube-reinforced composite plates using higher-order shear deformation theory, advancing numerical modeling techniques for complex composite structures. Rafiee et al. [15] studied the nonlinear dynamic stability of piezoelectric functionally graded carbon nanotube-reinforced composite plates with initial geometric imperfections, addressing the challenges associated with the dynamic behavior of smart composite structures. Shen [16] explored the nonlinear bending behavior of functionally graded carbon nanotube-reinforced composite plates in thermal environments, crucial for understanding their responses under varying temperature conditions. Thai and Vo [17] proposed a new sinusoidal shear deformation theory to analyze the bending, buckling, and vibration of functionally graded plates, providing a comprehensive framework for studying the mechanical behavior of these structures. Wattanasakulpong and Chaikittiratana [18] presented exact solutions for the static and dynamic analyses of carbon nanotube-reinforced composite plates with a Pasternak elastic foundation, contributing to the fundamental understanding of the mechanical behavior of composite materials. Ying et al. [19] investigated the stochastic micro-vibration response characteristics of sandwich plates with a magnetorheological (MR) visco-elastomer core and mass, providing valuable insights for designing vibration-damping systems in smart structures and systems.

However, to date, no report has been published in the literature on the dynamic stability of viscoelastic nanocomposite micro plates. Motivated by these considerations, in order to improve the optimum design of nanostructures, we aim to present a realistic model for the temperature-dependent dynamic instability of micro plates reinforced with FG-CNTs. The structural damping effects are assumed by the Kelvin–Voigt model. The surrounding elastic medium is simulated by orthotropic visco-Pasternak foundation. The mixture rule is applied for obtaining the equivalent material properties of the structure. The motion equations are obtained using the SSDT and energy method considering size effects. The DQM is used to calculate the resonance frequency and DIR of the structure. The effects of different parameters such as volume percent of CNTs, distribution type of CNTs, temperature, nonlocal parameters, and structural damping on the dynamic instability of the visco-system are elucidated.

8.2 MOTION EQUATIONS

As shown in Figure 8.1, a carbon nanotubes reinforced composite (CNTRC) visco-plate with length a, width b, and thickness h is considered. The CNTRC plate is

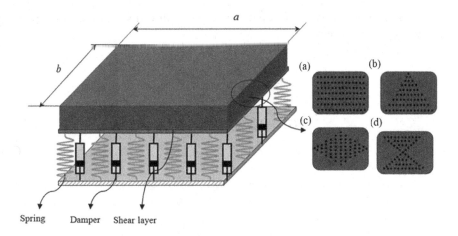

FIGURE 8.1 Configurations of the SWCNT distribution in a CNTRC plate. (a) UD-CNTRC plate; (b) FGA-CNTRC plate; (c) FGO-CNTRC plate; (d) FGX-CNTRC plate.

surrounded by an orthotropic elastomeric temperature-dependent medium, which is simulated by K_W, $K_{g\xi}$, and $K_{g\eta}$, corresponding to the Winkler foundation parameter and shear foundation parameters in the ξ and η directions, respectively. Four types of CNTRC plates, namely, uniform distribution (UD) along with three types of FG distributions (FGA, FGO, FGX) of CNTs along the thickness direction of a CNTRC plate, are considered.

In order to obtain the equivalent material properties of two-phase nanocomposites (i.e., polymer as the matrix and CNT as the reinforcer), the mixture rule is applied. According to the mixture rule, the effective Young's and shear moduli of the CNTRC plate can be written as [16]:

$$E_{11} = \eta_1 V_{CNT} E_{r11} + (1 - V_{CNT}) E_m, \tag{8.1}$$

$$\frac{\eta_2}{E_{22}} = \frac{V_{CNT}}{E_{r22}} + \frac{(1 - V_{CNT})}{E_m}, \tag{8.2}$$

$$\frac{\eta_3}{G_{12}} = \frac{V_{CNT}}{G_{r12}} + \frac{(1 - V_{CNT})}{G_m}, \tag{8.3}$$

where (E_{r1}, E_{r2}) and G_{r11} indicate the Young's moduli and shear modulus of SWCNTs, respectively, and E_m and G_m represent the corresponding properties of the isotropic matrix. The scale-dependent material properties, η_j $(j = 1, 2, 3)$, can be calculated by matching the effective properties of the CNTRC obtained from the MD simulations with those from the mixture rule. V_{CNT} and V_m are the volume fractions of the CNTs and matrix, respectively, the sum of which equals unity. The uniform and three types of FG distributions of the CNTs along the thickness direction of the CNTRC plates take the following forms:

$$UD: \quad V_{CNT} = V_{CNT}^*, \tag{8.4}$$

$$FGV: \quad V_{CNT}(z) = \left(1 - \frac{2z}{h}\right)V_{CNT}^*, \tag{8.5}$$

$$FGO: \quad V_{CNT}(z) = 2\left(1 - \frac{2|z|}{h}\right)V_{CNT}^*, \tag{8.6}$$

$$FGX: \quad V_{CNT}(z) = 2\left(\frac{2|z|}{h}\right)V_{CNT}^*, \tag{8.7}$$

where

$$V_{CNT}^* = \frac{w_{CNT}}{w_{CNT} + \left(\rho_{CNT}/\rho_m\right) - \left(\rho_{CNT}/\rho_m\right)w_{CNT}}, \tag{8.8}$$

where w_{CNT}, ρ_m, and ρ_{CNT} are the mass fraction of the CNTs and the densities of the matrix and CNTs, respectively. Similarly, the thermal expansion coefficients in the longitudinal and transverse directions (α_{11} and α_{22}, respectively) and the density (ρ) of the CNTRC plates can be determined as:

$$\rho = V_{CNT}^* \rho_r + V_m \rho_m, \tag{8.9}$$

$$\alpha_{11} = V_{CNT}^* \alpha_{r11} + V_m \alpha_m, \tag{8.10}$$

$$\alpha_{22} = \left(1 + \nu_{r12}\right)V_{CNT}\alpha_{r22} + \left(1 + \nu_m\right)V_m\alpha_m - \nu_{12}\alpha_{11}, \tag{8.11}$$

where ν_{r12} and ν_m are Poisson's ratios of the CNTs and matrix, respectively. In addition, (α_{r1}, α_{r22}) and α_m are the thermal expansion coefficients of the CNTs and matrix, respectively. It should be noted that ν_{12} is assumed constant over the thickness of the FG-CNTRC plates.

In Eringen's nonlocal elasticity model, the stress state at a reference point in the body is regarded to be dependent not only on the strain state at this point but also on the strain states at all of the points throughout the body. The constitutive equation of the nonlocal elasticity is [4]:

$$(1 - (e_0 a)^2 \nabla^2)\sigma_{ij} = C_{ijkl}\,\varepsilon_{kl}, \tag{8.12}$$

where the parameter e_0a denotes the small-scale parameter, and ∇^2 is the Laplace operator. The constitutive equation for the stress (σ) and strain (ε) matrix in the thermal environment may be written as follows:

$$\left(1-\left(e_0a\right)^2\nabla^2\right)\begin{Bmatrix}\sigma_{xx}\\ \sigma_{yy}\\ \sigma_{yz}\\ \sigma_{zx}\\ \sigma_{xy}\end{Bmatrix}=\begin{bmatrix}C_{11}\left(z,T\right) & C_{12}\left(z,T\right) & 0 & 0 & 0\\ C_{21}\left(z,T\right) & C_{22}\left(z,T\right) & 0 & 0 & 0\\ 0 & 0 & C_{44}\left(z,T\right) & 0 & 0\\ 0 & 0 & 0 & C_{55}\left(z,T\right) & 0\\ 0 & 0 & 0 & 0 & C_{66}\left(z,T\right)\end{bmatrix}$$

$$\begin{Bmatrix}\varepsilon_{xx}-\alpha_{11}\Delta T\\ \varepsilon_{yy}-\alpha_{22}\Delta T\\ \gamma_{yz}\\ \gamma_{xz}\\ \gamma_{xy}\end{Bmatrix},$$

(8.13)

where C_{ij} denotes the temperature-dependent elastic coefficients. Note that C_{ij} and (α_{11},α_{22}) may be obtained using the mixture rule. All materials exhibit some viscoelastic response. According to Kelvin and Voigt [8], in real life, nanostructure mechanical properties depend on the time variation. This model shows that as the stress is released, the material gradually relaxes to its undeformed state. By considering this model, we have:

$$C_{ij}=C_{ij}\left(1+g\frac{\partial}{\partial t}\right),$$

(8.14)

where g is the structural damping constant. Based on the sinusoidal plate theory presented in Chapter 1, the displacement field can be assumed. The potential energy of the structure can be written as:

$$U=\frac{1}{2}\int_A\int_{-\frac{h}{2}}^{\frac{h}{2}}\left(\sigma_{xx}\varepsilon_{xx}+\sigma_{yy}\varepsilon_{yy}+\sigma_{xy}\gamma_{xy}+\sigma_{xz}\gamma_{xz}+\sigma_{yz}\gamma_{yz}\right)dzdA$$

(8.15)

Substituting Eq. (8.13) into Eq. (8.15) leads to:

$$U=\frac{1}{2}\int_A\left(N_{xx}\frac{\partial u}{\partial x}+N_{xy}\frac{\partial u}{\partial y}+N_{xy}\frac{\partial v}{\partial x}+N_{yy}\frac{\partial v}{\partial y}+Q_x\frac{\partial w_s}{\partial x}+Q_y\frac{\partial w_s}{\partial y}-M_{xxS}\frac{\partial^2 w_s}{\partial x^2}\right.$$

$$\left.-M_{yyS}\frac{\partial^2 w_s}{\partial y^2}-2M_{xyS}\frac{\partial^2 w_s}{\partial y\partial x}-M_{xxB}\frac{\partial^2 w_b}{\partial x^2}-M_{yyB}\frac{\partial^2 w_b}{\partial y^2}-2M_{xyB}\frac{\partial^2 w_b}{\partial y\partial x}\right)dA,$$

(8.16)

where N, M, and Q are the stress resultant–displacement and can be defined by:

$$\left(N_{xx}, N_{yy}, N_{xy}\right) = \int_{-\frac{h}{2}}^{\frac{h}{2}} (\sigma_{xx}^{nl}, \sigma_{yy}^{nl}, \sigma_{xy}^{nl}) dz,$$ (8.17)

$$\left(M_{xxB}, M_{yyB}, M_{xyB}\right) = \int_{-\frac{h}{2}}^{\frac{h}{2}} (\sigma_{xz}^{nl}, \sigma_{yy}^{nl}, \sigma_{xy}^{nl}) z \, dz,$$ (8.18)

$$\left(M_{xxS}, M_{yyS}, M_{xyS}\right) = -\int_{-\frac{h}{2}}^{\frac{h}{2}} (\sigma_{xz}^{nl}, \sigma_{yy}^{nl}, \sigma_{xy}^{nl}) f \, dz,$$ (8.19)

$$\left(Q_x, Q_y\right) = \int_{-\frac{h}{2}}^{\frac{h}{2}} (\sigma_{xz}^{nl}, \sigma_{yz}^{nl}) g \, dz.$$ (8.20)

The kinetic energy of the nanocomposite micro plate can be written as:

$$K = \frac{1}{2} \rho \int_A \int_{-\frac{h}{2}}^{\frac{h}{2}} \left[\left(\frac{\partial U_1}{\partial t}\right)^2 + \left(\frac{\partial U_2}{\partial t}\right)^2 + \left(\frac{\partial U_3}{\partial t}\right)^2 \right] dz \, dA,$$ (8.21)

where ρ is the density of the structure. The external work due to the surrounding orthotropic visco-Pasternak medium can be written as:

$$W = -\int_A (q) u_3 \, dA,$$ (8.22)

where

$$q = kw + c_d \dot{w} - G_\xi \left(\cos^2 \theta w_{,xx} + 2\cos\theta\sin\theta w_{,yx} + \sin^2 \theta w_{,yy}\right)$$
$$- G_\eta \left(\sin^2 \theta w_{,xx} - 2\sin\theta\cos\theta w_{,yx} + \cos^2 \theta w_{,yy}\right),$$ (8.23)

where the angle θ describes the local ξ direction of the orthotropic foundation with respect to the global x-axis of the plate; k, G_ξ, and G_η are the Winkler foundation parameter and the shear foundation parameters in the ξ and η directions, respectively. Finally, applying Hamilton's principle, the motion equations can be obtained as:

$$\frac{\partial}{\partial x} N_{xx} + \frac{\partial}{\partial y} N_{xy} - I_0 \frac{\partial^2 U}{\partial t^2} + I_1 \frac{\partial^3 W_b}{\partial x \partial t^2} + J_1 \frac{\partial^3 W_s}{\partial x \partial t^2} = 0,$$ (8.24)

$$\frac{\partial}{\partial x} N_{xy} + \frac{\partial}{\partial y} N_{yy} - I_0 \frac{\partial^2 V}{\partial t^2} + I_1 \frac{\partial^3 W_b}{\partial y \partial t^2} + J_1 \frac{\partial^3 W_s}{\partial y \partial t^2} = 0,$$ (8.25)

$$\frac{\partial^2}{\partial r^2} M_{xxB} + 2\frac{\partial^2}{\partial x \partial y} M_{xyB} + \frac{\partial^2}{\partial y^2} M_{yyB} + q + N_x^m \left(\frac{\partial^2 W_b}{\partial x^2} + \frac{\partial^2 W_s}{\partial x^2} \right)$$

$$+ N_y^m \left(\frac{\partial^2 W_b}{\partial y^2} + \frac{\partial^2 W_s}{\partial y^2} \right) - I_0 \left(\frac{\partial^2 W_b}{\partial t^2} + \frac{\partial^2 W_s}{\partial t^2} \right) - I_1 \left(\frac{\partial^3 U}{\partial x \partial t^2} + \frac{\partial^3 V}{\partial y \partial t^2} \right)$$

$$+ I_2 \left(\frac{\partial^4 W_b}{\partial x^2 \partial t^2} + \frac{\partial^4 W_b}{\partial y^2 \partial t^2} \right) + J_2 \left(\frac{\partial^4 W_s}{\partial x^2 \partial t^2} + \frac{\partial^4 W_s}{\partial y^2 \partial t^2} \right) = 0, \tag{8.26}$$

$$\frac{\partial^2}{\partial x^2} M_{xxS} + 2\frac{\partial^2}{\partial x \partial y} M_{xyS} + \frac{\partial^2}{\partial y^2} M_{yyS} + \frac{\partial}{\partial x} Q_x + \frac{\partial}{\partial y} Q_y + q + + N_x^m \left(\frac{\partial^2 W_b}{\partial x^2} + \frac{\partial^2 W_s}{\partial x^2} \right)$$

$$+ N_y^m \left(\frac{\partial^2 W_b}{\partial y^2} + \frac{\partial^2 W_s}{\partial y^2} \right) - I_0 \left(\frac{\partial^2 W_b}{\partial t^2} + \frac{\partial^2 W_s}{\partial t^2} \right) - J_1 \left(\frac{\partial^3 U}{\partial x \partial t^2} + \frac{\partial^3 V}{\partial y \partial t^2} \right)$$

$$+ J_2 \left(\frac{\partial^4 W_b}{\partial x^2 \partial t^2} + \frac{\partial^4 W_b}{\partial y^2 \partial t^2} \right) + K_2 \left(\frac{\partial^4 W_s}{\partial x^2 \partial t^2} + \frac{\partial^4 W_s}{\partial y^2 \partial t^2} \right) = 0, \tag{8.27}$$

where the mass inertias can be defined as:

$$\left(I_0, I_1, I_2, J_1, J_1, K_2 \right) = \int_{-\frac{h}{2}}^{\frac{h}{2}} \rho \left(1, z, f, zf, z^2, f^2 \right) dz. \tag{8.28}$$

By substituting Eq. (8.13) into Eqs. (8.17)–(8.20) the stress resultants are obtained as:

$$N_{xx} = A_{11} \frac{\partial}{\partial x} U - A_{11z} \frac{\partial^2}{\partial x^2} W_b - A_{11f} \frac{\partial^2}{\partial x^2} W_s + A_{12} \frac{\partial}{\partial y} V - A_{12z} \frac{\partial^2}{\partial y^2} W_b - A_{12f} \frac{\partial^2}{\partial y^2} W_s +$$

$$G \left[\begin{array}{l} A_{11} \dfrac{\partial^2}{\partial x \partial t} U - A_{11z} \dfrac{\partial^3}{\partial x^2 \partial t} W_b - A_{11f} \dfrac{\partial^3}{\partial x^2 \partial t} W_s + A_{12} \dfrac{\partial^2}{\partial y \partial t} V \\ -A_{12z} \dfrac{\partial^3}{\partial y^2 \partial t} W_b - A_{12f} \dfrac{\partial^3}{\partial y^2 \partial t} W_s \end{array} \right], \tag{8.29}$$

$$N_{yy} = A_{21} \frac{\partial}{\partial x} U - A_{21z} \frac{\partial^2}{\partial x^2} W_b - A_{21f} \frac{\partial^2}{\partial x^2} W_s + A_{22} \frac{\partial}{\partial y} V - A_{22z} \frac{\partial^2}{\partial y^2} W_b - A_{22f} \frac{\partial^2}{\partial y^2} W_s +$$

$$G \left[\begin{array}{l} A_{12} \dfrac{\partial^2}{\partial x \partial t} U - A_{12z} \dfrac{\partial^3}{\partial x^2 \partial t} W_b - A_{12f} \dfrac{\partial^3}{\partial x^2 \partial t} W_s + A_{22} \dfrac{\partial^2}{\partial y \partial t} V \\ -A_{22z} \dfrac{\partial^3}{\partial y^2 \partial t} W_b - A_{22f} \dfrac{\partial^3}{\partial y^2 \partial t} W_s \end{array} \right], \tag{8.30}$$

$$N_{xy} = A_{44} \frac{\partial}{\partial y} U + A_{44} \frac{\partial}{\partial x} V - 2A_{44z} \frac{\partial^2}{\partial x \partial y} W_b - 2A_{44f} \frac{\partial^2}{\partial x \partial y} W_s$$

$$+ G \begin{bmatrix} A_{44} \dfrac{\partial^2}{\partial y \partial t} U + A_{44} \dfrac{\partial^2}{\partial x \partial t} V \\[2mm] -2A_{44z} \dfrac{\partial^3}{\partial x \partial y \partial t} W_b - 2A_{44f} \dfrac{\partial^3}{\partial x \partial y \partial t} W_s \end{bmatrix}, \tag{8.31}$$

$$Q_x = A_{55g} \frac{\partial}{\partial x} W_s + GA_{55g} \frac{\partial^2}{\partial x \partial t} W_s, \tag{8.32}$$

$$Q_y = A_{66g} \frac{\partial}{\partial y} W_s + GA_{66g} \frac{\partial^2}{\partial y \partial t} W_s, \tag{8.33}$$

$$M_{xxB} = A_{11z} \frac{\partial}{\partial x} U - B_{11} \frac{\partial^2}{\partial x^2} W_b - A_{11zf} \frac{\partial^2}{\partial x^2} W_s + A_{12z} \frac{\partial}{\partial y} V - B_{12} \frac{\partial^2}{\partial y^2} W_b - A_{12zf} \frac{\partial^2}{\partial y^2} W_s +$$

$$G \begin{bmatrix} A_{11z} \dfrac{\partial^2}{\partial x \partial t} U - B_{11} \dfrac{\partial^3}{\partial x^2 \partial t} W_b - A_{11zf} \dfrac{\partial^3}{\partial x^2 \partial t} W_s + A_{12z} \dfrac{\partial^2}{\partial y \partial t} V \\[2mm] - B_{12} \dfrac{\partial^3}{\partial y^2 \partial t} W_b - A_{12zf} \dfrac{\partial^3}{\partial y^2 \partial t} W_s \end{bmatrix}, \tag{8.34}$$

$$M_{xxS} = A_{11f} \frac{\partial}{\partial x} U - A_{11zf} \frac{\partial^2}{\partial x^2} W_b - E_{11} \frac{\partial^2}{\partial x^2} W_s + A_{12f} \frac{\partial}{\partial y} V - A_{12zf} \frac{\partial^2}{\partial y^2} W_b - E_{12} \frac{\partial^2}{\partial y^2} W_s +$$

$$G \begin{bmatrix} A_{11f} \dfrac{\partial^2}{\partial x \partial t} U - A_{11zf} \dfrac{\partial^3}{\partial x^2 \partial t} W_b - E_{11} \dfrac{\partial^3}{\partial x^2 \partial t} W_s + A_{12f} \dfrac{\partial^2}{\partial y \partial t} V \\[2mm] - A_{12f} \dfrac{\partial^3}{\partial y^2 \partial t} W_b - E_{12} \dfrac{\partial^3}{\partial y^2 \partial t} W_s \end{bmatrix}, \tag{8.35}$$

$$M_{yyB} = A_{21z} \frac{\partial}{\partial x} U - B_{21} \frac{\partial^2}{\partial x^2} W_b - A_{21zf} \frac{\partial^2}{\partial x^2} W_s + A_{22z} \frac{\partial}{\partial y} V - B_{22} \frac{\partial^2}{\partial y^2} W_b - A_{22zf} \frac{\partial^2}{\partial y^2} W_s +$$

$$G \begin{bmatrix} A_{21z} \dfrac{\partial^2}{\partial x \partial t} U - B_{21} \dfrac{\partial^3}{\partial x^2 \partial t} W_b - A_{21zf} \dfrac{\partial^3}{\partial x^2 \partial t} W_s + A_{22z} \dfrac{\partial^2}{\partial y \partial t} V \\[2mm] - B_{22} \dfrac{\partial^3}{\partial y^2 \partial t} W_b - A_{22zf} \dfrac{\partial^3}{\partial y^2 \partial t} W_s \end{bmatrix}, \tag{8.36}$$

$$M_{yyS} = A_{21f} \frac{\partial}{\partial x} U - A_{21zf} \frac{\partial^2}{\partial x^2} W_b - E_{21} \frac{\partial^2}{\partial x^2} W_s + A_{22f} \frac{\partial}{\partial y} V - A_{22zf} \frac{\partial^2}{\partial y^2} W_b - E_{22} \frac{\partial^2}{\partial y^2} W_s +$$

$$G \begin{bmatrix} A_{21f} \dfrac{\partial^2}{\partial x \partial t} U - A_{21zf} \dfrac{\partial^3}{\partial x^2 \partial t} W_b - E_{21} \dfrac{\partial^3}{\partial x^2 \partial t} W_s + A_{22f} \dfrac{\partial^2}{\partial y \partial t} V \\ -A_{22zf} \dfrac{\partial^3}{\partial y^2 \partial t} W_b - E_{22} \dfrac{\partial^3}{\partial y^2 \partial t} W_s \end{bmatrix}, \tag{8.37}$$

$$M_{xyB} = 2A_{44z} \frac{\partial}{\partial y} U + 2A_{44z} \frac{\partial}{\partial x} V - 2B_{44} \frac{\partial^2}{\partial x \partial y} W_b - 2A_{44zf} \frac{\partial^2}{\partial x \partial y} W_s +$$

$$G \left[2A_{44z} \frac{\partial^2}{\partial y \partial t} U + 2A_{44z} \frac{\partial^2}{\partial x \partial t} V - 2B_{44} \frac{\partial^3}{\partial x \partial y \partial t} W_b - 2A_{44zf} \frac{\partial^3}{\partial x \partial y \partial t} W_s \right], \tag{8.38}$$

$$M_{xyS} = 2A_{44f} \frac{\partial}{\partial y} U + 2A_{44f} \frac{\partial}{\partial x} V - 2A_{44zf} \frac{\partial^2}{\partial x \partial y} W_b - 2E_{44} \frac{\partial^2}{\partial x \partial y} W +$$

$$G \left[2A_{44f} \frac{\partial^2}{\partial y \partial t} U + 2A_{44f} \frac{\partial^2}{\partial x \partial t} V - 2A_{44zf} \frac{\partial^3}{\partial x \partial y \partial t} W_b - 2E_{44} \frac{\partial^3}{\partial x \partial y \partial t} W_s \right], \tag{8.39}$$

where

$$\left(A_{11}, A_{12}, A_{22}, A_{44} \right) = \sum_{k=1}^{N} \int_{z^{(k-1)}}^{z^{(k)}} \left(C_{11}^{(k)}, C_{12}^{(k)}, C_{22}^{(k)}, C_{44}^{(k)} \right) dz, \tag{8.40}$$

$$\left(A_{11z}, A_{12z}, A_{22z}, A_{44z} \right) = \sum_{k=1}^{N} \int_{z^{(k-1)}}^{z^{(k)}} \left(C_{11}^{(k)}, C_{12}^{(k)}, C_{22}^{(k)}, C_{44}^{(k)} \right) z \, dz, \tag{8.41}$$

$$\left(A_{11f}, A_{12f}, A_{22f}, A_{44f} \right) = \sum_{k=1}^{N} \int_{z^{(k-1)}}^{z^{(k)}} \left(C_{11}^{(k)}, C_{12}^{(k)}, C_{22}^{(k)}, C_{44}^{(k)} \right) f \, dz, \tag{8.42}$$

$$\left(A_{11zf}, A_{12zf}, A_{22zf}, A_{44zf} \right) = \sum_{k=1}^{N} \int_{z^{(k-1)}}^{z^{(k)}} \left(C_{11}^{(k)}, C_{12}^{(k)}, C_{22}^{(k)}, C_{44}^{(k)} \right) z f \, dz, \tag{8.43}$$

$$\left(A_{55g}, A_{66g} \right) = \sum_{k=1}^{N} \int_{z^{(k-1)}}^{z^{(k)}} \left(C_{55}^{(k)}, C_{66}^{(k)} \right) dz, \tag{8.44}$$

$$\left(B_{11}, B_{12}, B_{22}, B_{44} \right) = \sum_{k=1}^{N} \int_{z^{(k-1)}}^{z^{(k)}} \left(C_{11}^{(k)}, C_{12}^{(k)}, C_{22}^{(k)}, C_{44}^{(k)} \right) z^2 \, dz, \tag{8.45}$$

$$\left(E_{11}, E_{12}, E_{22}, E_{44}\right) = \sum_{k=1}^{N} \int_{z^{(k-1)}}^{z^{(k)}} \left(C_{11}^{(k)}, C_{12}^{(k)}, C_{22}^{(k)}, C_{44}^{(k)}\right) f^2 dz. \tag{8.46}$$

By substituting Eqs. (8.29)–(8.39) into Eqs. (8.24)–(8.27), the equations of motion can be expressed as:

$$
\begin{aligned}
&\frac{\partial}{\partial x}\left\{
\begin{aligned}
&A_{11}\frac{\partial}{\partial x}U - A_{11z}\frac{\partial^2}{\partial x^2}W_b - A_{11f}\frac{\partial^2}{\partial x^2}W_s + A_{12}\frac{\partial}{\partial y}V - A_{12z}\frac{\partial^2}{\partial y^2}W_b - A_{12f}\frac{\partial^2}{\partial y^2}W_s + \\
&G\left[
\begin{aligned}
&A_{11}\frac{\partial^2}{\partial x\partial t}U - A_{11z}\frac{\partial^3}{\partial x^2\partial t}W_b - A_{11f}\frac{\partial^3}{\partial x^2\partial t}W_s + A_{12}\frac{\partial^2}{\partial y\partial t}V \\
&-A_{12z}\frac{\partial^3}{\partial y^2\partial t}W_b - A_{12f}\frac{\partial^3}{\partial y^2\partial t}W_s
\end{aligned}
\right]
\end{aligned}
\right\} \\
&+\frac{\partial}{\partial y}\left\{
\begin{aligned}
&A_{44}\frac{\partial}{\partial y}U + A_{44}\frac{\partial}{\partial x}V - 2A_{44z}\frac{\partial^2}{\partial x\partial y}W_b - 2A_{44f}\frac{\partial^2}{\partial x\partial y}W \\
&+G\left[A_{44}\frac{\partial^2}{\partial y\partial t}U + A_{44}\frac{\partial^2}{\partial x\partial t}V - 2A_{44z}\frac{\partial^3}{\partial x\partial y\partial t}W_b - 2A_{44f}\frac{\partial^3}{\partial x\partial y\partial t}W_s\right]
\end{aligned}
\right\} \\
&=\left(1-\alpha\nabla^2\right)\left[I_0\frac{\partial^2 U}{\partial t^2} - I_1\frac{\partial^3 W_b}{\partial x\partial t^2} - J_1\frac{\partial^3 W_s}{\partial x\partial t^2}\right],
\end{aligned}
\tag{8.47}
$$

$$
\begin{aligned}
&\frac{\partial}{\partial x}\left\{
\begin{aligned}
&A_{44}\frac{\partial}{\partial y}U + A_{44}\frac{\partial}{\partial x}V - 2A_{44z}\frac{\partial^2}{\partial x\partial y}W - 2A_{44f}\frac{\partial^2}{\partial x\partial y}W_b \\
&+G\left[A_{44}\frac{\partial^2}{\partial y\partial t}U + A_{44}\frac{\partial^2}{\partial x\partial t}V - 2A_{44z}\frac{\partial^3}{\partial x\partial y\partial t}W_b - 2A_{44f}\frac{\partial^3}{\partial x\partial y\partial t}W_s\right]
\end{aligned}
\right\} \\
&+\frac{\partial}{\partial y}\left\{
\begin{aligned}
&A_{21}\frac{\partial}{\partial x}U - A_{21z}\frac{\partial^2}{\partial x^2}W_b - A_{21f}\frac{\partial^2}{\partial x^2}W_s + A_{22}\frac{\partial}{\partial y}V - A_{22z}\frac{\partial^2}{\partial y^2}W_b - A_{22f}\frac{\partial^2}{\partial y^2}W_s + \\
&G\left[
\begin{aligned}
&A_{12}\frac{\partial^2}{\partial x\partial t}U - A_{12z}\frac{\partial^3}{\partial x^2\partial t}W_b - A_{12f}\frac{\partial^3}{\partial x^2\partial t}W_s + A_{22}\frac{\partial^2}{\partial y\partial t}V \\
&-A_{22z}\frac{\partial^3}{\partial y^2\partial t}W_b - A_{22f}\frac{\partial^3}{\partial y^2\partial t}W_s
\end{aligned}
\right]
\end{aligned}
\right\} \\
&=\left(1-\alpha\nabla^2\right)\left[I_0\frac{\partial^2 V}{\partial t^2} - I_1\frac{\partial^3 W_b}{\partial y\partial t^2} - J_1\frac{\partial^3 W_s}{\partial y\partial t^2}\right],
\end{aligned}
\tag{8.48}
$$

$$\frac{\partial^2}{\partial x^2}\left[\begin{array}{l}\left(A_{11z}\dfrac{\partial}{\partial r}U - B_{11}\dfrac{\partial^2}{\partial x^2}W_b - A_{11zf}\dfrac{\partial^2}{\partial x^2}W_s + A_{12z}\dfrac{\partial}{\partial y}V - B_{12}\dfrac{\partial^2}{\partial y^2}W_b - A_{12zf}\dfrac{\partial^2}{\partial y^2}W_s + \right. \\ G\left[\begin{array}{l}A_{11z}\dfrac{\partial^2}{\partial x\partial t}U - B_{11}\dfrac{\partial^3}{\partial x^2\partial t}W_b - A_{11zf}\dfrac{\partial^3}{\partial x^2\partial t}W_s + A_{12z}\dfrac{\partial^2}{\partial y\partial t}V \\ -B_{12}\dfrac{\partial^3}{\partial y^2\partial t}W_b - A_{12zf}\dfrac{\partial^3}{\partial y^2\partial t}W_s\end{array}\right]\end{array}\right]$$

$$+2\frac{\partial^2}{\partial x\partial y}\left[\begin{array}{l}2A_{44z}\dfrac{\partial}{\partial y}U + 2A_{44z}\dfrac{\partial}{\partial x}V - 2B_{44}\dfrac{\partial^2}{\partial x\partial y}W_b - 2A_{44zf}\dfrac{\partial^2}{\partial x\partial y}W_s \\ +G\left[2A_{44z}\dfrac{\partial^2}{\partial y\partial t}U + 2A_{44z}\dfrac{\partial^2}{\partial x\partial t}V - 2B_{44}\dfrac{\partial^3}{\partial x\partial y\partial t}W_b - 2A_{44zf}\dfrac{\partial^3}{\partial x\partial y\partial t}W_s\right]\end{array}\right]$$

$$+\frac{\partial^2}{\partial y^2}\left[\begin{array}{l}\left(A_{21z}\dfrac{\partial}{\partial x}U - B_{21}\dfrac{\partial^2}{\partial x^2}W_b - A_{21zf}\dfrac{\partial^2}{\partial x^2}W_s + A_{22z}\dfrac{\partial}{\partial y}V - B_{22}\dfrac{\partial^2}{\partial y^2}W_b - A_{22zf}\dfrac{\partial^2}{\partial y^2}W_s + \right. \\ G\left[\begin{array}{l}A_{21z}\dfrac{\partial^2}{\partial x\partial t}U - B_{21}\dfrac{\partial^3}{\partial x^2\partial t}W_b - A_{21zf}\dfrac{\partial^3}{\partial x^2\partial t}W_s + A_{22z}\dfrac{\partial^2}{\partial y\partial t}V \\ -B_{22}\dfrac{\partial^3}{\partial y^2\partial t}W_b - A_{22zf}\dfrac{\partial^3}{\partial y^2\partial t}W_s\end{array}\right]\end{array}\right]$$

$$\left(1-\mu\nabla^2\right)\left[-kw - c_d\dot{w} + G_\xi\left(\cos^2\theta w_{,xx} + 2\cos\theta\sin\theta w_{,yx} + \sin^2\theta w_{,yy}\right)\right.$$

$$+G_\eta\left(\sin^2\theta w_{,xx} - 2\sin\theta\cos\theta w_{,yx} + \cos^2\theta w_{,yy}\right)$$

$$+N_x^m\left(\frac{\partial^2 W_b}{\partial x^2} + \frac{\partial^2 W_s}{\partial x^2}\right) + N_y^m\left(\frac{\partial^2 W_b}{\partial y^2} + \frac{\partial^2 W_s}{\partial y^2}\right) = I_0\left(\frac{\partial^2 W_b}{\partial t^2} + \frac{\partial^2 W_s}{\partial t^2}\right) + I_1\left(\frac{\partial^3 U}{\partial x\partial t^2} + \frac{\partial^3 V}{\partial y\partial t^2}\right)$$

$$-I_2\left(\frac{\partial^4 W_b}{\partial x^2\partial t^2} + \frac{\partial^4 W_b}{\partial y^2\partial t^2}\right) - J_2\left(\frac{\partial^4 W_s}{\partial x^2\partial t^2} + \frac{\partial^4 W_s}{\partial y^2\partial t^2}\right)\right], \tag{8.49}$$

$$
\frac{\partial^2}{\partial x^2}\left\{
\begin{array}{l}
A_{11f}\dfrac{\partial}{\partial x}U - A_{11zf}\dfrac{\partial^2}{\partial x^2}W_b - E_{11}\dfrac{\partial^2}{\partial x^2}W_s + A_{12f}\dfrac{\partial}{\partial y}V - A_{12zf}\dfrac{\partial^2}{\partial y^2}W_b - E_{12}\dfrac{\partial^2}{\partial y^2}W_s + \\[4mm]
G\left[
\begin{array}{l}
A_{11f}\dfrac{\partial^2}{\partial x\partial t}U - A_{11zf}\dfrac{\partial^3}{\partial x^2\partial t}W_b - E_{11}\dfrac{\partial^3}{\partial x^2\partial t}W_s + A_{12f}\dfrac{\partial^2}{\partial y\partial t}V \\[4mm]
-A_{12f}\dfrac{\partial^3}{\partial y^2\partial t}W_b - E_{12}\dfrac{\partial^3}{\partial y^2\partial t}W_s
\end{array}
\right]
\end{array}
\right\}
$$

$$
+2\frac{\partial^2}{\partial x\partial y}\left\{
\begin{array}{l}
2A_{44f}\dfrac{\partial}{\partial y}U + 2A_{44f}\dfrac{\partial}{\partial x}V - 2A_{44zf}\dfrac{\partial^2}{\partial x\partial y}W_b - 2E_{44}\dfrac{\partial^2}{\partial x\partial y}W \\[4mm]
+G\left[
\begin{array}{l}
2A_{44f}\dfrac{\partial^2}{\partial y\partial t}U + 2A_{44f}\dfrac{\partial^2}{\partial x\partial t}V - 2A_{44zf}\dfrac{\partial^3}{\partial x\partial y\partial t}W_b \\[4mm]
-2E_{44}\dfrac{\partial^3}{\partial x\partial y\partial t}W_s
\end{array}
\right]
\end{array}
\right\}
$$

$$
+\frac{\partial^2}{\partial y^2}\left\{
\begin{array}{l}
A_{21f}\dfrac{\partial}{\partial x}U - A_{21zf}\dfrac{\partial^2}{\partial x^2}W_b - E_{21}\dfrac{\partial^2}{\partial x^2}W_s + A_{22f}\dfrac{\partial}{\partial y}V - A_{22zf}\dfrac{\partial^2}{\partial y^2}W_b - E_{22}\dfrac{\partial^2}{\partial y^2}W_s + \\[4mm]
G\left[
\begin{array}{l}
A_{21f}\dfrac{\partial^2}{\partial x\partial t}U - A_{21zf}\dfrac{\partial^3}{\partial x^2\partial t}W_b - E_{21}\dfrac{\partial^3}{\partial x^2\partial t}W_s + A_{22f}\dfrac{\partial^2}{\partial y\partial t}V \\[4mm]
-A_{22f}\dfrac{\partial^3}{\partial y^2\partial t}W_b - E_{22}\dfrac{\partial^3}{\partial y^2\partial t}W_s
\end{array}
\right]
\end{array}
\right\}
$$

$$
+\frac{\partial}{\partial x}\left(A_{55g}\frac{\partial}{\partial x}W_s + GA_{55g}\frac{\partial^2}{\partial x\partial t}W_s\right) + \frac{\partial}{\partial y}\left(A_{66g}\frac{\partial}{\partial y}W_s + GA_{66g}\frac{\partial^2}{\partial y\partial t}W_s\right) +
$$

$$
\left(1-\mu\nabla^2\right)\left[-kw - c_d\dot{w} + G_\xi\left(\cos^2\theta w_{,xx} + 2\cos\theta\sin\theta w_{,yx} + \sin^2\theta w_{,yy}\right)\right.
$$

$$
+G_\eta\left(\sin^2\theta w_{,xx} - 2\sin\theta\cos\theta w_{,yx} + \cos^2\theta w_{,yy}\right)
$$

$$
+N_x^m\left(\frac{\partial^2 W_b}{\partial x^2} + \frac{\partial^2 W_s}{\partial x^2}\right) + N_y^m\left(\frac{\partial^2 W_b}{\partial y^2} + \frac{\partial^2 W_s}{\partial y^2}\right) = I_0\left(\frac{\partial^2 W_b}{\partial t^2} + \frac{\partial^2 W_s}{\partial t^2}\right)
$$

$$
\left.+J_1\left(\frac{\partial^3 U}{\partial x\partial t^2} + \frac{\partial^3 V}{\partial y\partial t^2}\right) - J_2\left(\frac{\partial^4 W_b}{\partial x^2\partial t^2} + \frac{\partial^4 W_b}{\partial y^2\partial t^2}\right) - K_2\left(\frac{\partial^4 W_s}{\partial x^2\partial t^2} + \frac{\partial^4 W_s}{\partial y^2\partial t^2}\right)\right]. \tag{8.50}
$$

Let the in-plane load P be periodic, and it may be expressed as:

$$
P(t) = \alpha P_{cr} + \beta P_{cr}\cos(\omega t), \tag{8.51}
$$

where ω is the frequency of excitation, P_{cr} is the static buckling load, and α and β may be defined as static and dynamic load factors, respectively. However, the motion equations can be written as based on the DQM presented in Chapter 2:

$$
\left\{[K - \alpha P_{cr}K_G - \beta P_{cr}\cos(\omega t)K_G][d] + [C][\dot{d}] + [M][\ddot{d}]\right\} = [0]. \tag{8.52}
$$

In order to determinate the boundaries of the dynamic instability regions, the method suggested by Bolotin is applied. Hence, the components of $\{d\}$ can be written in the Fourier series with the period $2T$ as [14]:

$$\{d\} = \sum_{k=1,3,\dots}^{\infty} \left[\{a\}_k \sin\frac{k\omega t}{2} + \{b\}_k \cos\frac{k\omega t}{2} \right]. \qquad (8.53)$$

According to this method, the first instability region is usually the most important in studies of structures. This is due to the fact that the first DIR is wider than the other DIRs, and structural damping in higher regions is neutralized. Substituting Eq. (8.53) into Eq. (8.52) and setting the coefficients of each sine and cosine as well as the sum of the constant terms to zero yields:

$$\left| \left([K_L + K_{NL}] - P_{cr}\alpha[K]_G \pm P_{cr}\frac{\beta}{2}[K]_G \mp [C]\frac{\omega}{2} - [M]\frac{\omega^2}{4} \right) \right| = 0, \qquad (8.54)$$

Solving this equation based on the eigenvalue problem, the variation in ω with respect to α can be plotted as the DIR.

8.3 NUMERICAL RESULTS

A computer program is prepared for the numerical solution of nonlinear buckling of CNTRC sandwich micro plates resting on an orthotropic elastomeric temperature-dependent foundation. Here, poly methyl methacrylate (PMMA) is selected for the matrix, which has a constant Poisson's ratio of $\nu_m = 0.34$, temperature-dependent thermal coefficient of $\alpha_m = (1 + 0.0005\Delta T) \times 10^{-6} / K$, and temperature-dependent Young's modulus of $E_m = (3.52 - 0.0034T)\, GPa$, where $T = T_0 + \Delta T$ and $T_0 = 300\ K$ (room temperature). In addition, (10, 10) SWCNTs are selected as reinforcements with the material properties listed in Table 8.1 [16].

The elastomeric medium is made of poly dimethylsiloxane (PDMS), the temperature-dependent material properties of which are assumed to be $\nu_s = 0.48$ and $E_s = (3.22 - 0.0034T)\, GPa$, where $T = T_0 + \Delta T$ and $T_0 = 300\ K$ (room temperature).

TABLE 8.1

Temperature-Dependent Material Properties of (10, 10) SWCNTs
($L = 9.26$ nm, $R = 0.68$ nm, $h = 0.067$ nm, $v_{12}^{CNT} = 0.175$)

T (K)	E_{11}^{CNT} TPa	E_{22}^{CNT} TPa	G_{12}^{CNT} TPa	α_{12}^{CNT} $(10^{-6} / K)$	α_{22}^{CNT} $(10^{-6} / K)$
300	5.6466	7.0800	1.9445	3.4584	5.1682
500	5.5308	6.9348	1.9643	4.5361	5.0189
700	5.4744	6.8641	1.9644	4.6677	4.8943

To the best of the authors' knowledge, no published literature is available for viscoelastic nanocomposite sandwich micro plates based on the SSDT. Since no reference to such a work is found to date in the literature, its validation is not possible. However, in an attempt to validate this chapter as far as possible, a simplified analysis is carried out without considering the nonlocal parameter, orthotropic visco-Pasternak foundation, viscoelastic property, or SWCNTs as reinforcers. The present results are compared with the work of Lanhe et al. [7] based on first-order shear deformation theory (FSDT). Considering the same material properties reported by Lanhe et al. [7] and the dimensionless frequency as $\bar{\omega} = \omega(a/h)^2\sqrt{\rho_m(1-v^2)/E_m}$, the results of the comparison are shown in Figure 8.2. As can be seen, the present results are in good agreement with those of Lanhe et al. [7], indicating the validity of this chapter. Note that the slight difference between this chapter and that of Lanhe et al. [7] is due to the consideration of thermal load by Lanhe et al. [7].

The convergence and accuracy of the DQM in calculating the excitation frequency of the CNTRC sandwich micro plates is shown in Figure 8.3. The high rate of convergence of the method are quite evident, and it is found that 15 DQM grid points can yield accurate results.

In realizing the influence of CNTs as reinforcement, Figure 8.4 is plotted. This figure shows the effects of CNT volume fraction on the dimensionless excitation frequency $\left(\Omega = \omega(a/h)^2\sqrt{\rho_m(1-v^2)/E_m}\right)$ with respect to dynamic load amplitude. In

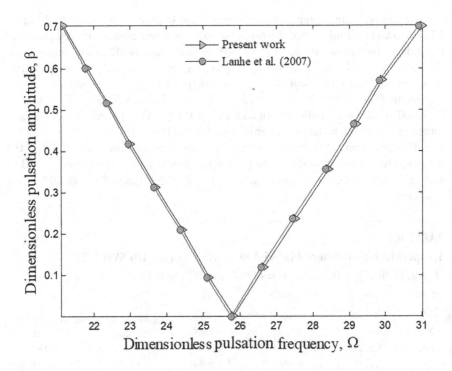

FIGURE 8.2 Comparison of the present work with Ref. [7].

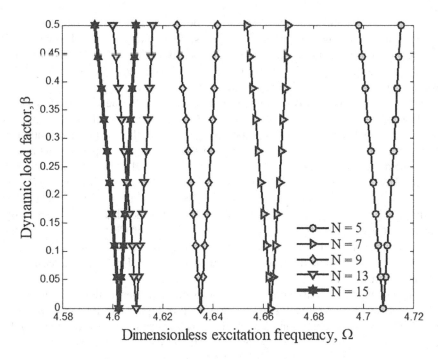

FIGURE 8.3 Convergence of the proposed method (DQM).

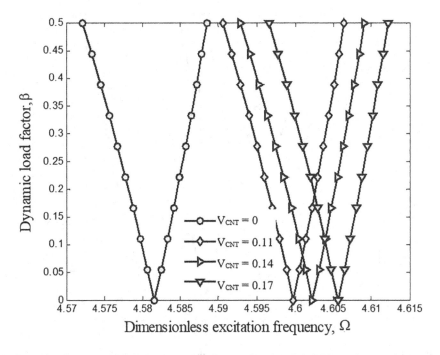

FIGURE 8.4 Effects of CNT volume percent on the DIR of the FG-CNT-reinforced micro plate.

this figure, the regions inside and outside the boundary curves correspond to unstable (parametric resonance) and stable regions, respectively. As can be seen, with an increasing CNT volume fraction, the DIR shifts to higher frequencies. In other words, increasing the CNT volume fraction leads to higher resonance frequency, which is due to an increase in the stiffness of the structure.

Depicted in Figure 8.5 is the dimensionless excitation frequency versus dynamic load factor for the UD and three types of FG-CNTRC sandwich micro plates. It should be noted that the mass fractions (w_{CNT}) of the UD and FG distribution of CNTs in the polymer are considered equal for the purpose of comparison. As can be seen, the dimensionless excitation frequencies of FGA- and FGO-CNTRC sandwich micro plates are lower than that of UD-CNTRC sandwich micro plate, while the FGX-CNTRC sandwich micro plate has a higher dimensionless excitation frequency with respect to the other three cases. This is due to the fact that the stiffness of the CNTRC sandwich micro plates changes with the form of CNT distribution in the matrix. However, it can be concluded that the CNTs distributed close to the top and bottom are more efficient than those distributed near the mid-plane for increasing the stiffness of the plates.

The effects of different boundary conditions on the dimensionless excitation frequency versus dynamic load factor are presented in Figure 8.6. It can be seen that the CCCC boundary condition yields a higher resonance frequency. In other words, comparing the assumed boundary conditions, the DIR of the structure moves to the right for the case of the micro plate with CCCC boundary conditions.

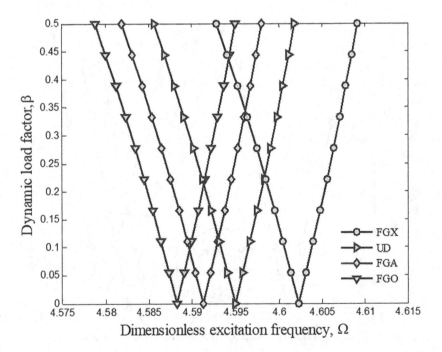

FIGURE 8.5 Effects of CNT distribution type on the DIR of FG-CNT-reinforced micro plate.

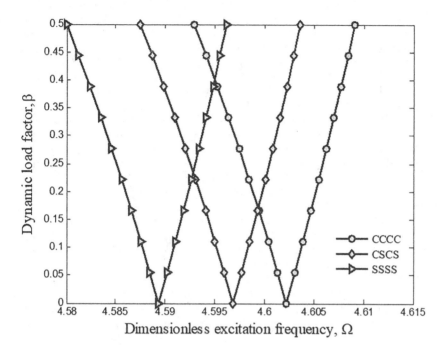

FIGURE 8.6 Effects of boundary condition on the DIR of the FG-CNT-reinforced micro plate.

Figure 8.7 demonstrates the effects of nonlocal parameters on the dimensionless excitation frequency versus dynamic load factor. As can be seen, the frequency of the system decreases when considering the nonlocal theory. This means that when considering the nonlocal parameter, the DIR of the system is observed at lower frequencies. This is because increasing the nonlocal parameter implies decreasing the interaction force between the micro plate atoms, leading to a softer structure.

Figure 8.8 demonstrates the DIRs for different structural damping constants. As can be seen, the DIR and frequency of the viscoelastic sandwich micro plate are lower than those of non-viscoelastic structure (i.e., $G = 0$). This remarkable difference shows that considering the nature of the nanocomposite micro plate as viscoelastic can yield more accurate results than assuming a non-viscoelastic micro plate. The reason is that assuming a viscoelastic micro plate means inducing a damping force, which results in more absorption of the vibration energy by the micro plate.

The effect of temperature on the dimensionless excitation frequency of the CNTRC sandwich micro plate with respect to the dynamic load factor is demonstrated in Figure 8.9. It can be seen that the dimensionless excitation frequency of the structure decreases with increasing temperature, which is due to the higher stiffness of the CNTRC micro plate with lower temperature.

The effect of different viscoelastic media is demonstrated in Figure 8.10 for four cases, which are with no viscoelastic medium, visco-Winkler medium, visco-Pasternak

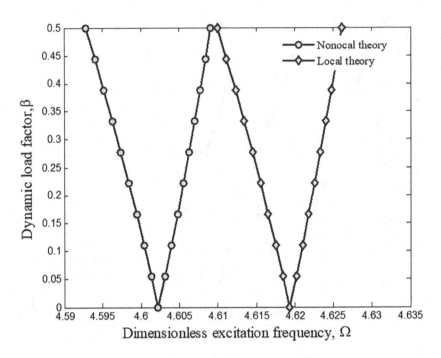

FIGURE 8.7 Effects of the nonlocal parameter on the DIR of the FG-CNT-reinforced micro plate.

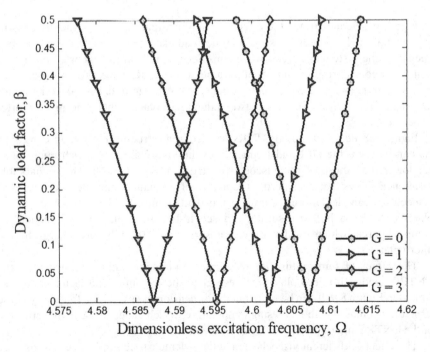

FIGURE 8.8 Effects of structural damping on the DIR of the FG-CNT-reinforced micro plate.

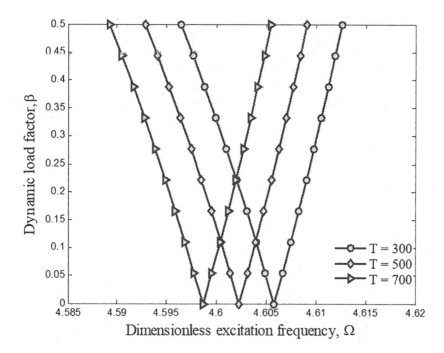

FIGURE 8.9 Effects of temperature on the DIR of the FG-CNT-reinforced micro plate.

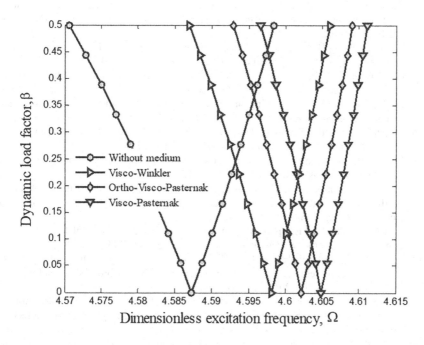

FIGURE 8.10 Effects of viscoelastic medium type on the DIR of the FG-CNT-reinforced micro plate.

medium, and orthotropic visco-Pasternak medium. It can be seen that considering the viscoelastic medium increases the excitation frequency of the structure and shifts the DIR to higher frequencies. This is due to the fact that considering the viscoelastic medium leads to a stiffer structure. Furthermore, the excitation frequency that leads to the DIR in the case of visco-Pasternak medium is higher than that for the visco-Winkler medium. This is because in the visco-Pasternak medium, the normal loads and the shear forces are considered. In addition, the DIR of the orthotropic visco-Pasternak medium is below the DIR of the visco-Pasternak medium since the shear layer is considered with the angle of 30°.

REFERENCES

1. Abdollahzadeh Shahrbabaki E, Alibeigloo A. Three-dimensional free vibration of carbon nanotube-reinforced composite plates with various boundary conditions using Ritz method. *Compos. Struct.* 2014;111:362–370.
2. Atteshamuddin SS, Yuwaraj MG. On the free vibration of angle-ply laminated composite and soft core sandwich plates. *Int. J. Sandw. Struct.* 2017 (in press).
3. Chetan SJ, Madhusudan M, Vidyashankar S, Charles Lu Y. Modelling of the interfacial damping due to nanotube agglomerations in nanocomposites. *Smart Struct. Syst.* 2017;19:57–66.
4. Eringen AC. On differential equations of nonlocal elasticity and solutions of screw dislocation and surface waves. *J. Appl. Phys.* 1983;544:703–710.
5. Ke LL, Liu C, Wang YS. Free vibration of nonlocal piezoelectric nanoplates under various boundary conditions. *Physica E.* 2015;66:93–106.
6. Kiani K. Free vibration of conducting nanoplates exposed to unidirectional in-plane magnetic fields using nonlocal shear deformable plate theories. *Physica E.* 2014;57:179–192.
7. Lanhe W, Wang H, Wang D. Dynamic stability analysis of FGM plates by the moving leastsquares differential quadrature method. *Compos. Struct.* 2007;77:383–394.
8. Lei Y, Adhikari S, Friswell MI. Vibration of nonlocal Kelvin–Voigt viscoelastic damped Timoshenko beams. *Int. J. Eng. Sci.* 2013;66–67:1–13.
9. Li YS, Cai ZY, Shi SY. Buckling and free vibration of magnetoelectroelastic nanoplate based on nonlocal theory. *Compos. Struct.* 2014;111:522–529.
10. Mantari JL, Guedes Soares C. Four-unknown quasi-3D shear deformation theory for advanced composite plates. *Compos. Struct.* 2014;109:231–239.
11. Mori T, Tanaka K. Average stress in matrix and average elastic energy of materials with misfitting inclusions. *Acta Metal. Mater.* 1973;21:571–574.
12. Natarajan S, Haboussi M, Manickam G. Application of higher-order structural theory to bending and free vibration analysis of sandwich plates with CNT reinforced composite facesheets. *Compos. Struct.* 2014;113:197–207.
13. Pandit MK, Sheikh AH, Singh BN. Analysis of laminated sandwich plates based on an improved higher order zigzag theory. *Int. J. Sandw. Struct.* 2009;12:307–326.
14. Phung-Van P, Abdel-Wahab M, Liew KM. Isogeometric analysis of functionally graded carbon nanotube-reinforced composite plates using higher-order shear deformation theory. *Compos. Struct.* 2017;123:137–149.
15. Rafiee M, He XQ, Liew KM. Non-linear dynamic stability of piezoelectric functionally graded carbon nanotube-reinforced composite plates with initial geometric imperfection. *Int. J. Non-Linear Mech.* 2014;59:37–51.
16. Shen HS. Nonlinear bending of functionally graded carbon nanotube-reinforced composite plates in thermal environments. *Compo. Struct.* 2009;91:9–19.

17. Thai HT, Vo TP. A new sinusoidal shear deformation theory for bending buckling and vibration of functionally graded plates. *Appl. Math. Model.* 2013;37:3269–3281.

18. Wattanasakulpong N, Chaikittiratana A. Exact solutions for static and dynamic analyses of carbon nanotube-reinforced composite plates with Pasternak elastic foundation. *Appl. Math. Model.* 2015;9:5459–5472.

19. Ying ZG, Ni YQ, Duan YF. Stochastic micro-vibration response characteristics of a sandwich plate with MR visco-elastomer core and mass. *Smart Struct. Syst.* 2017;16:141–162.

9 Stability of Nanocomposite Shells

9.1 INTRODUCTION

The mixture of fluid and nanoparticles produces a nanofluid. In nanofluids, common base nanoparticles include metals, oxides, carbides, or carbon nanotubes. The main advantage of nanofluids is heat resistance, which has many applications in fuel cells, hybrid-powered engines, chillers, heat exchangers, boilers, etc. Knowledge of the instability behavior of nanofluids in structures is found to be very critical in deciding their suitability for convective heat transfer applications.

The study of the nonlinear vibrations and stability of various structural elements, including shells, plates, and pipes, plays a crucial role in the design and analysis of engineering systems. This introduction provides an overview of recent research efforts focusing on the dynamic behavior and stability characteristics of fluid-conveying pipes, encompassing diverse material compositions, fluid flow conditions, and boundary conditions. Amabili [1] extensively reviewed the nonlinear vibrations and stability of shells and plates, offering valuable insights into the underlying mechanics governing these complex structures. Dai et al. [2] investigated the dynamics of fluid-conveying pipes composed of two different materials, highlighting the influence of material heterogeneity on structural response. Deng et al. [3] conducted a stability analysis of multi-span viscoelastic functionally graded material pipes conveying fluid, employing a hybrid method to capture the intricate interactions between material properties and fluid dynamics. Ghaitani and Majidian [4] explored the frequency and critical fluid velocity analysis of pipes reinforced with FG-CNTs, shedding light on the role of nanocomposite reinforcements in enhancing structural performance. He [5] proposed partitioned coupling strategies for fluid–structure interaction, presenting explicit, implicit, and semi-implicit schemes to address large displacement scenarios in fluid-conveying systems. Kutin and Bajsić [6] investigated the fluid-dynamic loading of pipes conveying fluid with laminar mean-flow velocity profiles, emphasizing the significance of flow regimes for structural behavior. Maalawi et al. [7] focused on the design of composite pipes conveying fluid to improve stability characteristics, contributing to the development of optimized structures for practical applications. Madani et al. [8] developed a differential cubature method for the vibration analysis of embedded FG-CNT-reinforced piezoelectric cylindrical shells subjected to uniform and non-uniform temperature distributions, addressing complex thermal effects on structural stability.

Marzani et al. [9] formulated finite element models for the dynamic instability analysis of fluid-conveying pipes on non-uniform elastic foundations, providing valuable tools for structural assessment in diverse environmental conditions.

DOI: 10.1201/9781003510710-9

Ni et al. [10] and Ni et al. [11] applied differential transformation methods to analyze the natural frequency and stability of pipes conveying fluid with axially moving supports immersed in fluid, offering efficient computational techniques for engineering analysis. Qian et al. [12] delved into the instability phenomenon of simply supported pipes conveying fluid under thermal loads, elucidating the intricate influence of thermal effects on the structural stability of such systems. By exploring the thermal-induced instabilities, their study underscored the critical role played by temperature variations in governing the structural response of fluid-conveying pipes, thus augmenting the understanding of thermal effects in engineering applications. Furthermore, theoretical advancements put forth by Reddy [13], Shen and Zhang [14], and Shokravi [15] have significantly contributed to the comprehension of composite materials and nonlocal beam models. These theoretical frameworks have provided comprehensive methodologies for analyzing the complex behaviors exhibited by structures composed of composite materials. By integrating advanced modeling techniques and theoretical insights, researchers can now better predict and understand the responses of composite structures under various loading conditions, enhancing the overall accuracy and reliability of structural analyses. Complementing theoretical developments, experimental and numerical investigations conducted by Rivero-Rodriguez and Pérez-Saborid [16], Ryu et al. [17], Sun and Gu [18], Texier and Dorbolo [19], and Wang [20] have yielded invaluable data pertaining to different aspects of fluid-conveying pipe dynamics. These studies have explored the influence of gravity, flutter instability, and distributed follower forces on the dynamic responses of fluid-conveying pipes through a combination of experimental measurements and numerical simulations. The insights gained from these investigations have contributed to refining theoretical models, validating numerical methods, and informing practical engineering designs aimed at improving the stability and performance of fluid-conveying pipe systems in various applications.

Based on the authors' knowledge, no report has been published on the instability of pipes conveying fluid–nanoparticle mixtures. However, the vibration and instability of embedded cylindrical shells conveying fluid-nanoparticles mixture have been studied. The material properties of a cylindrical shell and elastic foundation are assumed temperature dependent. Based on the FSDT, energy method, and Hamilton's principle, the motion equations are derived. Using the DQM, the frequency and critical fluid velocity of the structure are obtained. The effects of volume percent of nanoparticles, boundary conditions, geometrical parameters of the cylindrical shell, temperature change, elastic foundation, and fluid velocity on the frequency and critical fluid velocity of the structure are shown.

9.2 MOTION EQUATIONS

Figure 9.1 shows a cylindrical shell conveying fluid mixed with nanoparticles with the radius of R, thickness of h, length of L, and density of ρ. The elastic medium is modeled by the spring coefficient k_w and shear layer k_g.

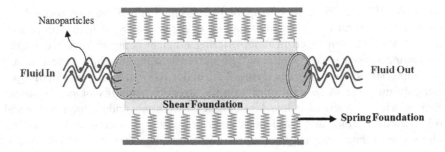

FIGURE 9.1 Scheme of a cylindrical shell conveying fluid mixed with nanoparticles surrounded by an elastic medium.

In this chapter, the first-order shell theory presented in Chapter 1 is used. The stress–strain relation can be written as:

$$\begin{Bmatrix} \sigma_{xx} \\ \sigma_{\theta\theta} \\ \sigma_{zz} \\ \tau_{\theta z} \\ \tau_{xz} \\ \tau_{x\theta} \end{Bmatrix} = \begin{bmatrix} C_{11} & C_{12} & C_{13} & 0 & 0 & 0 \\ C_{12} & C_{22} & C_{23} & 0 & 0 & 0 \\ C_{13} & C_{23} & C_{33} & 0 & 0 & 0 \\ 0 & 0 & 0 & C_{44} & 0 & 0 \\ 0 & 0 & 0 & 0 & C_{55} & 0 \\ 0 & 0 & 0 & 0 & 0 & C_{66} \end{bmatrix} \begin{Bmatrix} \varepsilon_{xx} - \alpha_{xx}T \\ \varepsilon_{\theta\theta} - \alpha_{\theta\theta}T \\ \varepsilon_{zz} - \alpha_{zz}T \\ \gamma_{\theta z} \\ \gamma_{xz} \\ \gamma_{x\theta} \end{Bmatrix}, \tag{9.1}$$

where $C_{ij}(i,j=1,2,..,6)$ denotes elastic coefficients; $\alpha_{xx}, \alpha_{\theta\theta}$ are thermal expansion; and T is the temperature rise, which follows a sinusoidal law as:

$$T(z) = \Delta T\left[1 - \cos\left[\frac{\pi}{2}\left(\frac{z}{h} + \frac{1}{2}\right)\right]\right] + T_i, \qquad \Delta T = T_o - T_i. \tag{9.2}$$

The potential energy of the structure is:

$$U = \frac{1}{2}\int \left(\sigma_{xx}\varepsilon_{xx} + \sigma_{\theta\theta}\varepsilon_{\theta\theta} + \tau_{xz}\gamma_{xz} + \tau_{\theta z}\gamma_{\theta z} + \tau_{x\theta}\gamma_{x\theta}\right)dV . \tag{9.3}$$

Replacing the strain–displacement equations in the previous relationships, we have:

$$U = \frac{1}{2}\int_0^{2\pi}\int_0^L \left\{ \left[N_{xx}\frac{\partial u}{\partial x} + M_{xx}\frac{\partial \phi_x}{\partial x}\right] + \left[\frac{N_{\theta\theta}}{R}\left(w + \frac{\partial v}{\partial \theta}\right) + \frac{M_{\theta\theta}}{R}\frac{\partial \phi_\theta}{\partial \theta}\right] + Q_x\left(\phi_x + \frac{\partial w}{\partial x}\right) \right. $$
$$\left. + \left(N_{x\theta}\left[\frac{\partial v}{\partial x} + \frac{1}{R}\frac{\partial u}{\partial \theta}\right] + M_{x\theta}\left[\frac{\partial \phi_\theta}{\partial x} + \frac{1}{R}\frac{\partial \phi_x}{\partial \theta}\right]\right) + Q_\theta\left[\frac{1}{R}\left(\frac{\partial w}{\partial \theta} - v\right) + \phi_\theta\right] \right\}Rdxd\theta, \tag{9.4}$$

where the stress resultant–displacement relations are as follows:

$$
\begin{bmatrix} N_x, M_x \\ N_\theta, M_\theta \\ N_{x\theta}, M_{x\theta} \end{bmatrix} = \int_{-\frac{h}{2}}^{\frac{h}{2}} \begin{bmatrix} \sigma_x \\ \sigma_\theta \\ \tau_{x\theta} \end{bmatrix} (1, z) dz,
\tag{9.5}
$$

$$
\begin{bmatrix} Q_x \\ Q_\theta \end{bmatrix} = k' \int_{-\frac{h}{2}}^{\frac{h}{2}} \begin{bmatrix} \tau_{xz} \\ \tau_{\theta z} \end{bmatrix} dz,
\tag{9.6}
$$

in which k' is the shear correction coefficient. The kinetic energy of the structure can be written as:

$$
K = \frac{\rho}{2} \int \left[\left(\frac{\partial u}{\partial t} + z \frac{\partial \varphi_x}{\partial t} \right)^2 + \left(\frac{\partial v}{\partial t} + z \frac{\partial \varphi_\theta}{\partial t} \right)^2 + \left(\frac{\partial w}{\partial t} \right)^2 \right] dV,
\tag{9.7}
$$

where ρ is the density of the structure. The moments of inertia are defined as follows:

$$
\begin{bmatrix} I_0 \\ I_1 \\ I_2 \end{bmatrix} = \int_{-h/2}^{h/2} \begin{bmatrix} \rho \\ \rho z \\ \rho z^2 \end{bmatrix} dz,
\tag{9.8}
$$

and the kinetic energy may be written as:

$$
K = \frac{1}{2} \int \begin{pmatrix} I_0 \left[\left(\frac{\partial u}{\partial t} \right)^2 + \left(\frac{\partial v}{\partial t} \right)^2 + \left(\frac{\partial w}{\partial t} \right)^2 \right] + I_1 \left(2 \frac{\partial u}{\partial t} \frac{\partial \varphi_x}{\partial t} + 2 \frac{\partial v}{\partial t} \frac{\partial \varphi_\theta}{\partial t} \right) \\ + I_2 \left[\left(\frac{\partial \varphi_x}{\partial t} \right)^2 + \left(\frac{\partial \varphi_\theta}{\partial t} \right)^2 \right] \end{pmatrix} dA.
\tag{9.9}
$$

The external work is due to the elastic foundation and fluid. The external work due to the elastic medium is [15]:

$$
W_e = -\int \left(k_w w - k_g \nabla^2 w \right) w dA,
\tag{9.10}
$$

in which k_w and k_g are the spring and shear constants, respectively. The external work caused by the fluid pressure is written as [20]:

$$
W_f = \int (F_{fluid}) w dA = \int \left(-\rho_E A \left(\frac{\partial^2 w}{\partial t^2} + 2 v_x \frac{\partial^2 w}{\partial x \partial t} + v_x^2 \frac{\partial^2 w}{\partial x^2} \right) \right.
$$
$$
\left. + \mu A \left(\frac{\partial^3 w}{\partial x^2 \partial t} + \frac{\partial^3 w}{R^2 \partial \theta^2 \partial t} + v_x \left(\frac{\partial^3 w}{\partial x^3} + \frac{\partial^3 w}{R^2 \partial \theta^2 \partial x} \right) \right) \right) w dA,
\tag{9.11}
$$

where $\propto$, ρ_E, and v_x are the viscosity, density, and velocity of the fluid–nanoparticle mixture, respectively. The fluid density and viscosity can be obtained using the mixture rule considering nanoparticles in the fluid as follows:

$$\rho_f = \gamma \rho_{np} + (1-\gamma)\rho_f, \tag{9.12}$$

$$\mu = (1+7.3\gamma+123\gamma^2)\mu_f, \tag{9.13}$$

where ρ_f and ρ_{np} are respectively the density of the fluid and nanoparticles, μ_{np} and μ_f are respectively the viscosity of the fluid and nanoparticles, and γ is the volume percent of the nanoparticles in the fluid. Applying Hamilton's principle, the motion equations can be written as:

$$\delta u : \frac{\partial N_{xx}}{\partial x} + \frac{\partial N_{x\theta}}{R\partial \theta} = I_0 \frac{\partial^2 u}{\partial t^2} + I_1 \frac{\partial^2 \varphi_x}{\partial t^2}, \tag{9.14}$$

$$\delta v : \frac{\partial N_{x\theta}}{\partial x} + \frac{\partial N_{\theta\theta}}{R\partial \theta} + \frac{Q_\theta}{R} = I_0 \frac{\partial^2 v}{\partial t^2} + I_1 \frac{\partial^2 \varphi_\theta}{\partial t^2}, \tag{9.15}$$

$$\delta w : \frac{\partial Q_x}{\partial x} + \frac{\partial Q_\theta}{R\partial \theta} - \frac{N_{\theta\theta}}{R} - k_w w + k_g \nabla^2 w - \rho_f A \left(\frac{\partial^2 w}{\partial t^2} + 2v_x \frac{\partial^2 w}{\partial x \partial t} + v_x^2 \frac{\partial^2 w}{\partial x^2} \right)$$
$$+ \mu A \left(\frac{\partial^3 w}{\partial x^2 \partial t} + \frac{\partial^3 w}{R^2 \partial \theta^2 \partial t} + v_x \left(\frac{\partial^3 w}{\partial x^3} + \frac{\partial^3 w}{R^2 \partial \theta^2 \partial x} \right) \right) = I_0 \frac{\partial^2 w}{\partial t^2}, \tag{9.16}$$

$$\delta \varphi_x : \frac{\partial M_{xx}}{\partial x} + \frac{\partial M_{x\theta}}{R\partial \theta} - Q_x = I_2 \frac{\partial^2 \varphi_x}{\partial t^2} + I_1 \frac{\partial^2 u}{\partial t^2}, \tag{9.17}$$

$$\delta \varphi_\theta : \frac{\partial M_{x\theta}}{\partial x} + \frac{\partial M_{\theta\theta}}{R\partial \theta} - Q_\theta = I_2 \frac{\partial^2 \varphi_\theta}{\partial t^2} + I_1 \frac{\partial^2 v}{\partial t^2}. \tag{9.18}$$

By integrating Eqs. (9.5) and (9.6) in the direction of thickness, the relationships of the forces and interior moments of the structure can be calculated as:

$$N_{xx} = A_{110} \frac{\partial u}{\partial x} + A_{111} \frac{\partial \phi_x}{\partial x} + A_{120} \left(\frac{\partial v}{R\partial \theta} + \frac{w}{R} \right) + A_{121} \frac{\partial \phi_\theta}{R\partial \theta}, \tag{9.19}$$

$$N_{\theta\theta} = A_{120} \frac{\partial u}{\partial x} + A_{121} \frac{\partial \phi_x}{\partial x} + A_{220} \left(\frac{\partial v}{R\partial \theta} + \frac{w}{R} \right) + A_{221} \frac{\partial \phi_\theta}{R\partial \theta}, \tag{9.20}$$

$$Q_\theta = k' A_{44} \left[\frac{1}{R} \left(\frac{\partial w}{\partial \theta} - v \right) + \phi_\theta \right], \tag{9.21}$$

$$Q_x = k' A_{55} \left(\frac{\partial w}{\partial x} + \phi_x \right), \tag{9.22}$$

$$N_{x\theta} = A_{660}\left(\frac{\partial u}{R\partial\theta} + \frac{\partial v}{\partial x}\right) + A_{661}\left(\frac{\partial\phi_x}{R\partial\theta} + \frac{\partial\phi_\theta}{\partial x}\right), \quad (9.23)$$

$$M_{xx} = A_{111}\frac{\partial u}{\partial x} + A_{112}\frac{\partial\phi_x}{\partial x} + A_{121}\left(\frac{\partial v}{R\partial\theta} + \frac{w}{R}\right) + A_{122}\frac{\partial\phi_\theta}{R\partial\theta}, \quad (9.24)$$

$$M_{\theta\theta} = A_{121}\frac{\partial u}{\partial x} + A_{122}\frac{\partial\phi_x}{\partial x} + A_{221}\left(\frac{\partial v}{R\partial\theta} + \frac{w}{R}\right) + A_{222}\frac{\partial\phi_\theta}{R\partial\theta}, \quad (9.25)$$

$$M_{x\theta} = A_{661}\left(\frac{\partial u}{R\partial\theta} + \frac{\partial v}{\partial x}\right) + A_{662}\left(\frac{\partial\varphi_x}{R\partial\theta} + \frac{\partial\varphi_\theta}{\partial x}\right), \quad (9.26)$$

where

$$\left(A_{11k}, A_{12k}, A_{22k}, A_{66k}\right) = \int_{-h/2}^{h/2}\left(C_{11}, C_{12}, C_{22}, C_{66}\right)z^k dz, \ k = 0,1,2 \quad (9.27)$$

$$\left(A_{44}, A_{55}\right) = \int_{-h/2}^{h/2}\left(C_{44}, C_{55}\right)dz. \quad (9.28)$$

In this chapter, three types of boundary conditions are used, which are:

- **Simple–Simple (SS)**
  $$x = 0, L \Rightarrow u = v = w = \phi_\theta = M_x = 0, \quad (9.29)$$

- **Clamped–Clamped (CC)**
  $$x = 0, L \Rightarrow u = v = w = \phi_x = \phi_\theta = 0, \quad (9.30)$$

- **Clamped–Simple (CS)**
  $$x = 0 \Rightarrow u = v = w = \phi_x = \phi_\theta = 0,$$
  $$x = L \Rightarrow u = v = w = \phi_x = M_x = 0. \quad (9.31)$$

Using the following time modes, the terms with the time derivative are omitted, and the differential equations will be entirely based on the local derivatives:

$$d(x,y,t) = d_0(x,y)e^{\omega t}, \quad (9.32)$$

where ω refers to frequency, and d shows the dynamic vector. Hence, the governing equations can be written as follows in a matrix form based on the DQM (please see Chapter 2):

$$\left(\underbrace{\left[K_L + K_{NL}\right]}_{K} + \Omega[C] + \Omega^2[M]\right)\begin{Bmatrix}\{d_b\}\\\{d_d\}\end{Bmatrix} = 0, \quad (9.33)$$

in which $\Omega = \omega/h\sqrt{C_{11}/\rho}$ refers to the dimensionless frequency. $[K_L], [K_{NL}], [C]$ and, $[M]$ show the linear part of the stiffness matrix, the nonlinear part of the stiffness matrix, the damper matrix, and the mass matrix, respectively. $\{d_b\}$ and $\{d_d\}$ are the dynamic range vectors in points of the boundary and domain. Based on an eigenvalue problem, Eq. (9.33) can be written as

$$
\begin{bmatrix} [0] & [I] \\ -[M^{-1}K] & -[M^{-1}C] \end{bmatrix} \{Z\} = \Omega\{Z\},
\tag{9.34}
$$

in which $[I]$ shows the identity matrix and $[0]$ is the zero matrix.

9.3 NUMERICAL RESULTS

In this section, the numerical results of the pipe conveying a fluid–nanoparticle mixture are presented. The pipe is made of poly methyl methacrylate (PMMA) for the matrix, which has a constant Poisson's ratio of $\nu_m = 0.34$, temperature-dependent thermal coefficient of $\alpha_m = (1 + 0.0005\Delta T) \times 10^{-6} / K$, and temperature-dependent Young's modulus of $E_m = (3.52 - 0.0034T)\ GPa$, where $T = T_0 + \Delta T$ and $T_0 = 300\ K$ (room temperature). The density of the fluid (water) is $\rho_f = 998.2\ Kg / m^3$, and its viscosity is $\alpha_f = 1 \times 10^{-3}\ Pa.s$. The nanoparticles in the fluid are iron oxide with a density of $\rho_{np} = 3970\ Kg / m^3$. Since the surrounding medium is relatively soft, the foundation stiffness k_w may be expressed by [14]:

$$
k_w = \frac{E_0}{4L(1 - \nu_0^2)(2 - c_1)^2} \left[5 - \left(2\gamma_1^2 + 6\gamma_1 + 5\right)\exp(-2\gamma_1)\right],
\tag{9.35}
$$

where

$$
c_1 = (\gamma_1 + 2)\exp(-\gamma_1),
\tag{9.36}
$$

$$
\gamma_1 = \frac{H_s}{L},
\tag{9.37}
$$

$$
E_0 = \frac{E_s}{(1 - \nu_s^2)},
\tag{9.38}
$$

$$
\nu_0 = \frac{\nu_s}{(1 - \nu_s)},
\tag{9.39}
$$

where E_s, ν_s, and H_s are the Young's modulus, Poisson's ratio, and depth of the foundation, respectively. In this chapter, E_s is assumed to be temperature dependent, while ν_s is assumed to be constant. The elastomeric medium is made of poly dimethylsiloxane (PDMS), the temperature-dependent material properties of which are assumed to be $\nu_s = 0.48$ and $E_s = (3.22 - 0.0034T)\ GPa$, where $T = T_0 + \Delta T$ and $T_0 = 300\ K$ (room temperature) [14].

In order to show the accuracy of the present work, neglecting the elastic medium and nanoparticles in the fluid, the results are compared with the work of Amabili [1]. However, considering a pipe with an elastic modulus of $E = 206\ GPa$, Poisson's ratio of $\nu = 0.3$, density $\rho = 7850\ Kg\,/\,m^3$, length-to-radius ratio of $L\,/\,R = 2$, and thickness-to-radius ratio of $h\,/\,R = 0.01$, the dimensionless frequency $\left(\Omega = \omega\,/\,\left\{\pi^2\,/\,L^2\left[D\,/\,\rho h\right]\right\}^{0.5}\right)$ is plotted versus the dimensionless fluid velocity $\left(V = v_x\,/\,\left\{\pi^2\,/\,L\left[D\,/\,\rho h\right]\right\}^{0.5}\right)$ in Figures 9.2 and 9.3. As can be seen, the present results are agreeing well with the results of Amabili [1].

Figure 9.4 shows the variation in dimensionless frequency $\left(\Omega = \omega\,/\,\left\{\pi^2\,/\,L^2\left[E_m\,/\,\rho_m\right]\right\}^{0.5}\right)$ versus grid point number for different dimensionless fluid velocities $\left(Vx = v_x\,/\,\left\{\pi^2\,/\,L\left[E_m\,/\,\rho_m\right]\right\}^{0.5}\right)$. It can be seen that the dimensionless frequency is decreased with increasing the grid point number, and for $N = 15$, the results converge.

In general, in all of the following figures, with increasing fluid velocity, the frequency of the structure decreases until reaching zero. In this state, the critical fluid velocity is observed. After the critical fluid velocity, the real part of the frequency has two values of positive and negative, where the positive one makes the structure divergence instable.

Figures 9.5 and 9.6 illustrate the effects of nanoparticle volume percent on the dimensionless frequency and damping of the structure, respectively. A direct relationship can be seen between nanoparticle volume percent and frequency of the structure so that with an increasing nanoparticle volume percent, the dimensionless

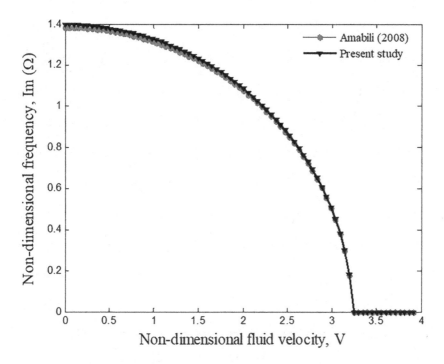

FIGURE 9.2 Validation of frequency.

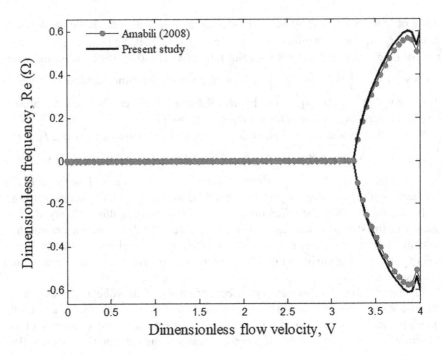

FIGURE 9.3 Validation of damping.

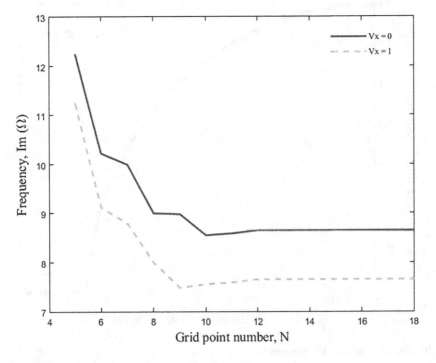

FIGURE 9.4 Convergence and accuracy of the DQM.

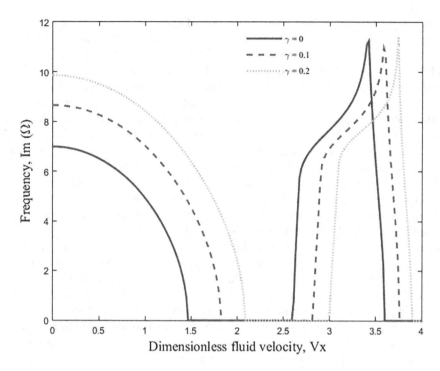

FIGURE 9.5 The effect of volume percent of nanoparticles on the frequency.

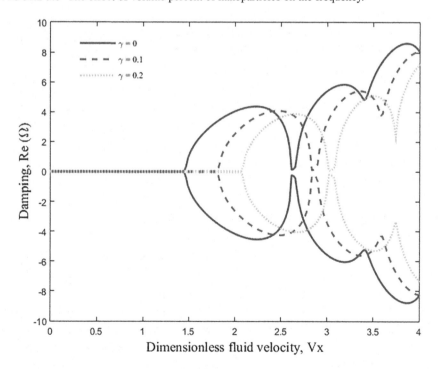

FIGURE 9.6 The effect of volume percent of nanoparticles on the damping.

frequency and critical fluid velocity increase. This is because with an increasing nanoparticle volume percent, the fluid velocity that leads to instability is reduced.

The effects of the elastic medium on the dimensionless frequency and damping of the pipe versus the dimensionless fluid velocity are shown in Figures 9.7 and 9.8, respectively. Three cases with no elastic medium, Winkler medium, and Pasternak medium are considered. It can be seen that considering the elastic medium raises the dimensionless frequency and critical fluid velocity of the structure due to an increase in the stiffness of the system. In addition, the dimensionless frequency and critical fluid velocity of the cylindrical shell located in Pasternak medium are higher than those for the shell surrounded by a Winkler foundation. This is due to the fact that the Pasternak medium considers two elements of normal and shear forces.

Figures 9.9 and 9.10 present the effects of different boundary conditions on the dimensionless frequency and damping of the pipe versus the dimensionless fluid velocity, respectively. It can be observed that the clamped–clamped (CC) boundary condition leads to higher dimensionless frequency and critical fluid velocity with respect to other considered boundary conditions. This is due to the fact that the cylindrical shell with the CC boundary condition has higher bending rigidity.

The effects of temperature change on the dimensionless frequency and damping of the pipe versus the dimensionless fluid velocity are shown in Figures 9.11 and 9.12, respectively. It can be concluded that with an increasing temperature change, the stiffness of the cylindrical shell is reduced, and consequently, the dimensionless frequency and critical fluid velocity are decreased.

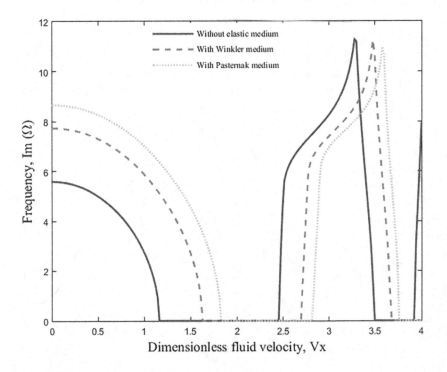

FIGURE 9.7 The effect of the elastic foundation on the frequency.

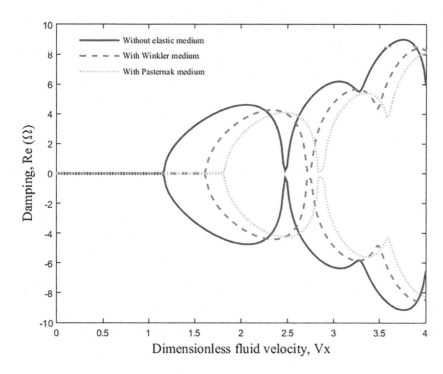

FIGURE 9.8 The effect of the elastic foundation on the damping.

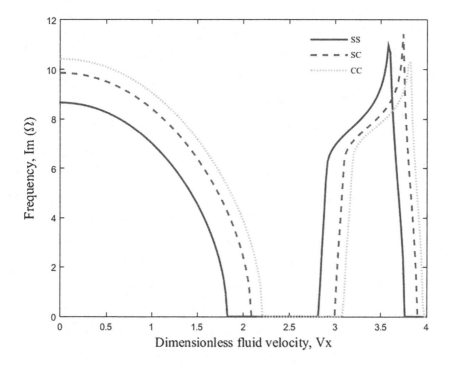

FIGURE 9.9 The effect of different boundary conditions on the frequency.

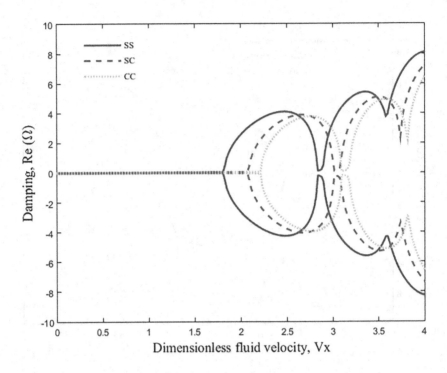

FIGURE 9.10 The effect of different boundary conditions on the damping.

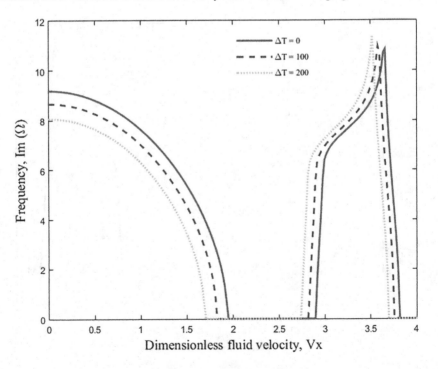

FIGURE 9.11 The effect of temperature change on the frequency.

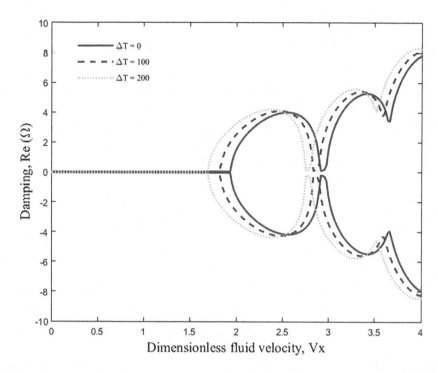

FIGURE 9.12 The effect of temperature change on the damping.

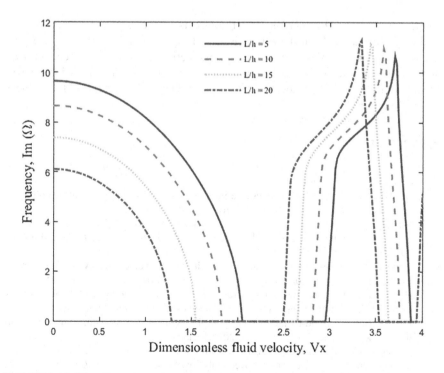

FIGURE 9.13 The effect of length-to-thickness ratio of the cylindrical shell on the frequency.

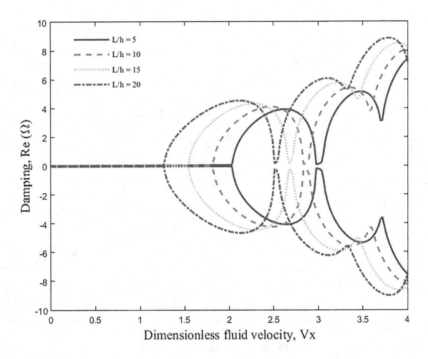

FIGURE 9.14 The effect of length-to-thickness ratio of the cylindrical shell on the damping.

Figures 9.13 and 9.14 demonstrate the effects of the length-to-thickness ratio of the cylindrical shell on the dimensionless frequency and damping of the pipe against the dimensionless fluid velocity, respectively. As can be seen, with an increasing length-to-thickness ratio of the cylindrical shell, the dimensionless frequency and critical fluid velocity are decreased due to a reduction in the stiffness of the structure.

REFERENCES

1. Amabili M. *Nonlinear vibrations and stability of shells and plates*. New York: Cambridge University Press, 2008.
2. Dai HL, Wang L, Ni Q. Dynamics of a fluid-conveying pipe composed of two different materials. *Int. J. Eng. Sci.* 2013;73:67–76.
3. Deng J, Liu Y, Zhang Z, Liu W. Stability analysis of multi-span viscoelastic functionally graded material pipes conveying fluid using a hybrid method. *Europ. J. Mech. A/Solids.* 2017;65:257–270.
4. Ghaitani MM, Majidian A. Frequency and critical fluid velocity analysis of pipes reinforced with FG-CNTs conveying internal flows. *Wind Struct.* 2017;24:267–285.
5. He T. Partitioned coupling strategies for fluid-structure interaction with large displacement: Explicit, implicit and semi-implicit schemes. *Wind Struct.* 2015;20:423–448.
6. Kutin J, Bajsić I. Fluid-dynamic loading of pipes conveying fluid with a laminar mean-flow velocity profile. *J. Fluids Struct.* 2014;50:171–183.
7. Maalawi KY, Abouel-Fotouh AM, Bayoumi ME, Yehia KAA. Design of composite pipes conveying fluid for improved stability characteristics. *Int. J. Appl. Eng. Res.* 2016;11:7633–7639.

8. Madani H, Hosseini H, Shokravi M. Differential cubature method for vibration analysis of embedded FG-CNT-reinforced piezoelectric cylindrical shells subjected to uniform and non-uniform temperature distributions. *Steel Compos. Struct.* 2017;22:889–913.

9. Marzani A, Mazzotti M, Viola E, Vittori P, Elishakoff I. FEM formulation for dynamic instability of fluid-conveying pipe on nonuniform elastic foundation. *Mech. Based Des. Struct. Mach.* 2012;40:83–95.

10. Ni Q, Zhang ZL, Wang L. Application of the differential transformation method to vibration analysis of pipes conveying fluid. *Appl. Math. Comput.* 2011;217:7028–7038.

11. Ni Q, Luo Y, Li M, Yan H. Natural frequency and stability analysis of a pipe conveying fluid with axially moving supports immersed in fluid. *J. Sound Vib.* 2017;403:173–189.

12. Qian Q, Wang L, Ni Q. Instability of simply supported pipes conveying fluid under thermal loads. *Mech. Res. Commun.* 2009;36:413–417.

13. Reddy JN. *Mechanics of laminated composite plates and shells.* 2nd edition. Washington: CRC Press, 2004.

14. Shen HSH, Zhang CHL. Nonlocal beam model for nonlinear analysis of carbon nanotubes on elastomeric substrates. *Comput. Mat. Sci.* 2011;50:1022–1029.

15. Shokravi M. Buckling analysis of embedded laminated plates with agglomerated CNT-reinforced composite layers using FSDT and DQM. *Geomech. Eng.* 2017;12:327–346.

16. Rivero-Rodriguez J, Pérez-Saborid M. Numerical investigation of the influence of gravity on flutter of cantilevered pipes conveying fluid. *J. Fluids Struct.* 2015;55:106–121.

17. Ryu BJ, Ryu SU, Kim GH, Yim KB. Vibration and dynamic stability of pipes conveying fluid on elastic foundations. *KSME Int. J.* 2004;18:2148–2157.

18. Sun FJ, Gu M. A numerical solution to fluid-structure interaction of membrane structures under wind action. *Wind Struct.* 2014;19:35–58.

19. Texier BD, Dorbolo S. Deformations of an elastic pipe submitted to gravity and internal fluid flow. *J. Fluids Struct.* 2015;55:364–371.

20. Wang L. Flutter instability of supported pipes conveying fluid subjected to distributed follower forces. *Acta. Mech. Solida. Sin.* 2012;25:46–52.

21. Shokravi M. Vibration analysis of silica nanoparticles-reinforced concrete beams considering agglomeration effects. *Comput. Concr.* 2017;19:333–338.

10 Vibration of Nanocomposite Shells

10.1 INTRODUCTION

A great deal of interest in the analysis of carbon nanotube-reinforced composite (CNTRC) structures is being manifested in the specialized literature. This interest is mainly due to the advent of new composite material systems exhibiting exotic properties as compared to the traditional carbon fiber-reinforced composite structures. Due to their very attractive thermomechanical properties, these new materials are going to play a great role in the construction of microelectromechanical systems (MEMS) and nanoelectromechanical systems (NEMS).

In the domain of structural mechanics and materials science, extensive research has been conducted to explore various aspects of composite structures, including their static and dynamic behavior, vibration characteristics, and active control mechanisms. A plethora of studies have investigated different types of composite materials, such as functionally graded materials (FGMs), carbon nanotube (CNT)-reinforced composites, and piezoelectric bonded smart structures, aiming to enhance their mechanical properties, durability, and performance under various loading conditions.

Alibeigloo and Nouri [1] explored the static analysis of functionally graded cylindrical shells with piezoelectric layers using the differential quadrature method. Alibeigloo [2] extended this research by investigating the static analysis of functionally graded carbon nanotube-reinforced composite plates embedded in piezoelectric layers. Bodaghi and Shakeri [3] proposed an analytical approach for the free vibration and transient response of functionally graded piezoelectric cylindrical panels subjected to impulsive loads. Jinhua et al. [4] focused on the dynamic stability of piezoelectric laminated cylindrical shells with delamination, shedding light on the structural response of such configurations. Nanda and Nath [5] delved into the realm of active control by studying delaminated composite shells with piezoelectric sensor/actuator patches, aiming to mitigate vibration-induced damage. Lei et al. [6] and Lei et al. [7] contributed to the dynamic stability analysis of carbon nanotube-reinforced functionally graded cylindrical panels and laminated FG-CNT-reinforced composite rectangular plates using advanced numerical techniques. Similarly, Lei et al. [8] and Lei et al. [9] investigated the elastodynamic and buckling behavior of CNT-reinforced functionally graded plates using element-free and meshless approaches. Liew et al. [10] investigated the postbuckling behavior of carbon nanotube-reinforced functionally graded cylindrical panels under axial compression, employing a meshless approach. Liu et al. [11] explored the thermo-electromechanical vibration of piezoelectric nanoplates based on nonlocal theory, providing insights into their unique dynamic characteristics. Rajesh

DOI: 10.1201/9781003510710-10

Bhangale and Ganesan [12] conducted free vibration studies of non-homogeneous functionally graded magnetoelectroelastic finite cylindrical shells, contributing to the understanding of their vibrational behavior. Shen [13] and Shen [14] investigated the postbuckling and nonlinear bending of functionally graded composite plates with piezoelectric actuators under various loading conditions. Sheng and Wang [15] and Sheng and Wang [16] focused on active control and dynamic behavior studies of functionally graded laminated cylindrical shells with PZT layers under different loading scenarios. Additionally, Sheng and Wang [17] delved into the thermoelastic vibration and buckling analysis of functionally graded piezoelectric cylindrical shells, offering valuable insights into their thermal response. Heng and Wang [18] conducted studies on the dynamic behavior of functionally graded cylindrical shells with PZT layers under moving loads, shedding light on the structural response of such configurations to dynamic forces. Additionally, they investigated the thermoelastic vibration and buckling analysis of functionally graded piezoelectric cylindrical shells [19], contributing valuable insights into the thermal behavior of these structures. Tiersten [20] provided foundational knowledge on linear piezoelectric plate vibrations, serving as a fundamental reference for further research in this area. Yaqoob Yasin and Kapuria [21] developed an efficient finite element formulation for smart piezoelectric composite and sandwich shallow shells, facilitating accurate analysis and design optimization of such structures.

In their investigations of functionally graded plates, Zhang and colleagues meticulously examined the thermal responses of these structures, scrutinizing the effects of temperature gradients on their stability and performance [22]. Furthermore, their studies elucidated the intricate interplay between mechanical deformation and thermal effects, providing valuable insights into the behavior of functionally graded materials under thermal loading scenarios.

Moreover, Zhang et al. delved into the realm of carbon nanotube (CNT)-reinforced cylindrical panels, where they conducted comprehensive analyses to elucidate the complex mechanical behavior exhibited by these composite structures. Through their research, they explored the phenomenon of large deformation, shedding light on the structural response of CNT-reinforced panels subjected to significant mechanical loads [23]. Additionally, their investigations extended to the realm of vibration control, where they proposed innovative strategies for actively mitigating vibrations in composite structures, thereby enhancing their dynamic performance [24]. In order to delve deeper into their research inquiries, Zhang and his collaborators harnessed state-of-the-art numerical methodologies, such as meshless methods and finite element analysis (FEA). These cutting-edge computational tools served as indispensable aids in their quest to unravel the complexities inherent in the mechanical behavior of composite structures [25–32]. Meshless methods, characterized by their grid-free nature, offered Zhang's team a flexible and powerful means of simulating the behavior of composite materials without the constraints imposed by traditional mesh-based approaches. By discretizing the domain into a set of particles or nodes and employing interpolation techniques, meshless methods enabled the researchers to accurately model the intricate geometries and material heterogeneities inherent in composite structures. This approach proved particularly valuable in scenarios involving large deformations or complex geometric configurations, where traditional mesh-based methods may encounter difficulties or limitations. In parallel, finite element analysis (FEA) provided Zhang and his team with another indispensable tool for their investigations. FEA is a widely used numerical technique for approximating the behavior of complex structures by discretizing them into smaller, more manageable elements. By solving the governing equations of motion for each element and assembling them into a system-level model, FEA allowed the researchers to simulate the mechanical response of composite structures under various loading conditions with high fidelity and precision. Through the synergistic utilization of meshless methods and FEA, Zhang and his collaborators were able to conduct detailed simulations and analyses of composite structures, shedding light on the underlying mechanisms governing their behavior. By systematically varying parameters such as material composition, loading conditions, and structural geometry, they gained invaluable insights into the fundamental principles governing the response of composite materials. This deeper understanding not only advanced the field of composite structural mechanics but also laid the groundwork for the development of more robust and efficient design methodologies for composite structures in diverse engineering applications. In their comprehensive exploration of composite structures, Zhang and his team delved into various aspects of vibration analysis and postbuckling behavior, significantly broadening the scope of their contributions to the field [33–45]. They embarked on an in-depth investigation into the dynamic response of functionally graded carbon nanotube-reinforced composite plates, considering intricate factors such as elastically restrained edges and internal column supports [33–38]. Their research endeavors encompassed a diverse array of topics, including the utilization of advanced computational techniques like the element-free method and finite element analysis. Through meticulous analysis and innovative methodologies, they conducted detailed studies on the postbuckling behavior of bi-axially compressed plates and laminated nanocomposite structures [36, 40]. Moreover, their research extended to the computation of aerothermoelastic properties and the optimization of shape control for composite plates, showcasing their holistic approach to addressing multifaceted challenges in composite structural mechanics [41, 42]. By employing these sophisticated computational tools, Zhang and his collaborators not only unraveled the underlying mechanisms governing the vibrational and postbuckling characteristics of composite materials but also laid the groundwork for the development of more efficient and reliable composite structures in engineering applications. Their contributions have significantly advanced the understanding of composite structural mechanics, offering invaluable insights for the design and optimization of composite materials in various engineering disciplines.

However, to date, no report has been found in the literature on the vibration analysis of FG-CNT-reinforced piezoelectric cylindrical shells subjected to non-uniform temperate distribution. Motivated by these considerations, in the present chapter, the temperature-dependent vibration analysis of an embedded FG-CNT-reinforced piezoelectric cylindrical shell is investigated. The structure is subjected to externally applied voltage with uniform and non-uniform temperate distributions. The mixture rule is used for calculating the equivalent material properties of the nanocomposite structure. The surrounding elastic medium is simulated by a Pasternak foundation. The motion equations are derived based on the FSDT in conjunction with Hamilton's principal. The frequency of the structure is obtained using the DCM for different boundary conditions. The effects of applied voltage, volume percent and distribution type of CNTs in the polymer, temperature distribution type, elastic medium, and boundary conditions on the frequency of the system are disused in detail.

10.2 STRESS–STRAIN RELATIONS

As shown in Figure 10.1, an FG-CNT-reinforced piezoelectric cylindrical shell with length L and thickness h is considered. The structure is surrounded by Pasternak medium. One uniform and three functionally graded CNT distributions are considered. However, for calculating the equivalent material properties of present nanocomposite structure, the mixture rule is applied.

Based on the first-order shell theory presented in Chapter 1, the deflection may be expressed. The subsequent characterization of electromechanical coupling covers the various classes of piezoelectric materials. Details with respect to definition and determination of the constants describing these materials have been standardized by the Institute of Electrical and Electronics Engineers. Stresses σ and strains ε on the

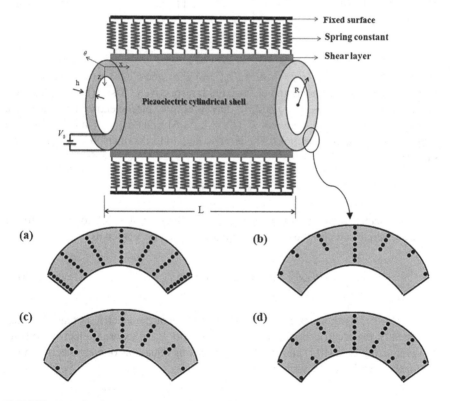

FIGURE 10.1 Configurations of the CNT distribution in a piezoelectric cylindrical shell. (a) UD; (b) FGV; (c) FGO; (d) FGX.

mechanical side, as well as flux density D and field strength E on the electrostatic side, may be arbitrarily combined as follows [20]:

$$
\begin{bmatrix} \sigma_{xx} \\ \sigma_{\theta\theta} \\ \sigma_{zz} \\ \tau_{\theta z} \\ \tau_{xz} \\ \tau_{x\theta} \end{bmatrix} = \begin{bmatrix} C_{11} & C_{12} & C_{13} & 0 & 0 & 0 \\ C_{12} & C_{22} & C_{23} & 0 & 0 & 0 \\ C_{13} & C_{23} & C_{33} & 0 & 0 & 0 \\ 0 & 0 & 0 & C_{44} & 0 & 0 \\ 0 & 0 & 0 & 0 & C_{55} & 0 \\ 0 & 0 & 0 & 0 & 0 & C_{66} \end{bmatrix} \left(\begin{Bmatrix} \varepsilon_{xx} - \alpha_{xx}\Delta T \\ \varepsilon_{\theta\theta} - \alpha_{\theta\theta}\Delta T \\ \varepsilon_{zz} - \alpha_{zz}\Delta T \\ \gamma_{\theta z} \\ \gamma_{xz} \\ \gamma_{x\theta} \end{Bmatrix} \right) - \begin{bmatrix} 0 & 0 & e_{31} \\ 0 & 0 & e_{32} \\ 0 & 0 & e_{33} \\ 0 & e_{24} & 0 \\ e_{15} & 0 & 0 \\ 0 & 0 & 0 \end{bmatrix} \begin{Bmatrix} E_x \\ E_\theta \\ E_z \end{Bmatrix},
$$

(10.1)

$$
\begin{bmatrix} D_x \\ D_\theta \\ D_z \end{bmatrix} = \begin{bmatrix} 0 & 0 & 0 & 0 & e_{15} & 0 \\ 0 & 0 & 0 & e_{24} & 0 & 0 \\ e_{31} & e_{31} & e_{33} & 0 & 0 & 0 \end{bmatrix} \left(\begin{Bmatrix} \varepsilon_{xx} - \alpha_{xx}\Delta T \\ \varepsilon_{\theta\theta} - \alpha_{\theta\theta}\Delta T \\ \varepsilon_{zz} - \alpha_{zz}\Delta T \\ \gamma_{\theta z} \\ \gamma_{xz} \\ \gamma_{x\theta} \end{Bmatrix} \right) + \begin{bmatrix} \in_{11} & 0 & 0 \\ 0 & \in_{22} & 0 \\ 0 & 0 & \in_{33} \end{bmatrix} \begin{Bmatrix} E_x \\ E_\theta \\ E_z \end{Bmatrix},
$$

(10.2)

where $\sigma_{ij}(i,j=x,\theta,z)$, $\varepsilon_{ij}(i,j=x,\theta,z)$, $D_{ii}(i=x,\theta,z)$, and $E_{ii}(i=x,\theta,z)$ are the stress, strain, electric displacement, and electric field, respectively. Also, $Q_{ij}(i,j=1,2,..,6)$, $e_{ij}(i,j=1,3,4,5)$, and $\in_{ij}(i,j=1,2,3)$ denote elastic, piezoelectric, and dielectric coefficients, respectively. Note that $C_{ij}(i,j=1,2,..,6)$ and $(\alpha_{xx},\alpha_{\theta\theta})$ may be obtained using mixture rule (next section). The electric field in terms of electric potential (Φ) is expressed as:

$$
E_k = -\nabla\Phi,
$$

(10.3)

where the electric potential is assumed as the combination of a half-cosine and linear variation, which satisfies the Maxwell equation. It can be written as [7]:

$$
\Phi(x,\theta,z,t) = -\cos(\frac{\pi z}{h})\varphi(x,\theta,t) + \frac{2V_0 z}{h},
$$

(10.4)

where $\varphi(x,\theta,t)$ is the time and spatial distribution of the electric potential, which must satisfy the electric boundary conditions, and V_0 is external electric voltage. However, using Eqs. (10.1) and (10.2), the governing equations of piezoelectric material for the FSDT may be written as:

$$
\sigma_{xx} = C_{11}\left(\varepsilon_{xx} - \alpha_{xx}\Delta T\right) + C_{12}\left(\varepsilon_{\theta\theta} - \alpha_{\theta\theta}\Delta T\right) + e_{31}\left(\frac{\pi}{h}\sin\left(\frac{\pi z}{h}\right)\varphi + \frac{2V_0}{h}\right),
$$

(10.5)

$$\sigma_{\theta\theta} = C_{12}\left(\varepsilon_{xx} - \alpha_{xx}\Delta T\right) + C_{22}\left(\varepsilon_{\theta\theta} - \alpha_{\theta\theta}\Delta T\right) + e_{32}\left(\frac{\pi}{h}\sin\left(\frac{\pi z}{h}\right)\varphi + \frac{2V_0}{h}\right), \quad (10.6)$$

$$\sigma_{\theta z} = C_{44}\varepsilon_{\theta z} - e_{15}\left(\cos\left(\frac{\pi z}{h}\right)\frac{\partial\varphi}{R\partial\theta}\right), \quad (10.7)$$

$$\sigma_{zx} = C_{55}\varepsilon_{zx} - e_{24}\left(\cos\left(\frac{\pi z}{h}\right)\frac{\partial\varphi}{\partial x}\right), \quad (10.8)$$

$$\sigma_{x\theta} = C_{66}\gamma_{x\theta}, \quad (10.9)$$

$$D_x = e_{15}\varepsilon_{xz} + \in_{11}\left(\cos\left(\frac{\pi z}{h}\right)\frac{\partial\varphi}{R\partial\theta}\right), \quad (10.10)$$

$$D_\theta = e_{24}\varepsilon_{z\theta} + \in_{22}\left(\cos\left(\frac{\pi z}{h}\right)\frac{\partial\varphi}{R\partial\theta}\right), \quad (10.11)$$

$$D_x = e_{31}\varepsilon_{xx} + e_{32}\varepsilon_{\theta\theta} - \in_{33}\left(\frac{\pi}{h}\sin\left(\frac{\pi z}{h}\right)\varphi + \frac{2V_0}{h}\right). \quad (10.12)$$

In order to accurately describe the effect of the temperature rise through the thickness, different temperature distributions are taken into account in the present analysis.

The cylindrical shell initial temperature is assumed to be T_i. The temperature is uniformly raised to a final value T. The temperature change is given by:

$$\Delta T = T - T_i. \quad (10.13)$$

The temperature of the outer surface is T_o, and it is considered to vary linearly from T_t to the inner surface temperature T_i. Therefore, the temperature rise through the thickness is given by:

$$T(z) = \Delta T\left(\frac{z}{h} + \frac{1}{2}\right) + T_i, \quad \Delta T = T_o - T_i. \quad (10.14)$$

In the third case, the temperature distribution across the thickness direction follows a sinusoidal law as:

$$T(z) = \Delta T\left[1 - \cos\left[\frac{\pi}{2}\left(\frac{z}{h} + \frac{1}{2}\right)\right]\right] + T_i, \quad \Delta T = T_o - T_i. \quad (10.15)$$

According to this theory, the effective Young's and shear moduli of the structure may be expressed as [33]:

$$E_{11} = \eta_1 V_{CNT} E_{r11} + (1 - V_{CNT}) E_m,$$ (10.16)

$$\frac{\eta_2}{E_{22}} = \frac{V_{CNT}}{E_{r22}} + \frac{(1 - V_{CNT})}{E_m},$$ (10.17)

$$\frac{\eta_3}{G_{12}} = \frac{V_{CNT}}{G_{r12}} + \frac{(1 - V_{CNT})}{G_m},$$ (10.18)

where (E_{r11}, E_{r22}) and E_m are the Young's moduli of CNTs and the matrix, respectively; G_{r11} and G_m are the shear moduli of CNTs and the matrix, respectively; V_{CNT} and V_m show the volume fractions of the CNTs and matrix, respectively; and η_j (j = 1, 2, 3) is the CNT efficiency parameter for considering the size-dependent material properties. Note that this parameter may be calculated using molecular dynamics (MD). However, the CNT distributions for the mentioned patterns obey the following [40]:

$$UD: \quad V_{CNT} = V_{CNT}^*,$$ (10.19)

$$FGV: \quad V_{CNT}(z) = \left(1 + \frac{2z}{h}\right) V_{CNT}^*,$$ (10.20)

$$FGO: \quad V_{CNT}(z) = 2\left(1 - \frac{2|z|}{h}\right) V_{CNT}^*,$$ (10.21)

$$FGX: \quad V_{CNT}(z) = 2\left(\frac{2|z|}{h}\right) V_{CNT}^*.$$ (10.22)

Furthermore, the thermal expansion coefficients in the axial and transverse directions (α_{11} and α_{22}, respectively) and the density (ρ) of the nanocomposite structure can be written as [10]:

$$\rho = V_{CNT} \rho_r + V_{m,} \rho_{m,},$$ (10.23)

$$\alpha_{11} = V_{CNT} \alpha_{r11} + V_{m,} \alpha_{m,},$$ (10.24)

$$\alpha_{22} = \left(1 + \nu_{r12}\right) V_{CNT} \alpha_{r22} + \left(1 + \nu_m\right) V_m \alpha_m - \nu_{12} \alpha_{11},$$ (10.25)

where

$$V_{CNT}^* = \frac{w_{CNT}}{w_{CNT} + (\rho_{CNT} / \rho_m) - (\rho_{CNT} / \rho_m) w_{CNT}}, \qquad (10.26)$$

where w_{CNT} is the mass fraction of the CNTs; ρ_m and ρ_{CNT} represent the densities of the matrix and CNTs, respectively; ν_{r12} and ν_m are Poisson's ratios of the CNTs and matrix, respectively; and (α_{r11}, α_{r22}) and α_m are the thermal expansion coefficients of the CNTs and matrix, respectively. Note that ν_{12} is assumed constant.

The total potential energy, V, of the system is the sum of potential energy, U, kinetic energy, K, and the work done by the elastic medium, W. The potential energy can be written as:

$$U = \frac{1}{2} \int \left(\begin{matrix} \sigma_{xx}\varepsilon_{xx} + \sigma_{\theta\theta}\varepsilon_{\theta\theta} + \sigma_{xz}\varepsilon_{xz} + \sigma_{\theta z}\varepsilon_{\theta z} + \sigma_{x\theta}\gamma_{x\theta} \\ -D_x E_x - D_\theta E_\theta - D_z E_z \end{matrix} \right) dV . \qquad (10.27)$$

Combining Eqs. (10.5)–(10.12) and (10.27) yields:

$$U = \frac{1}{2} \int_0^{2\pi} \int_0^L \left\{ \left[N_{xx} \frac{\partial u}{\partial x} + M_{xx} \frac{\partial \phi_x}{\partial x} \right] + \left[\frac{N_{\theta\theta}}{R} \left(w + \frac{\partial v}{\partial \theta} \right) + \frac{M_{\theta\theta}}{R} \frac{\partial \phi_\theta}{\partial \theta} \right] + Q_x \left(\phi_x + \frac{\partial w}{\partial x} \right) \right.$$

$$\left. + \left[N_{x\theta} \left(\frac{\partial v}{\partial x} + \frac{1}{R} \frac{\partial u}{\partial \theta} \right) + M_{x\theta} \left(\frac{\partial \phi_\theta}{\partial x} + \frac{1}{R} \frac{\partial \phi_x}{\partial \theta} \right) \right] + Q_\theta \left[\frac{1}{R} \left(\frac{\partial w}{\partial \theta} - v \right) + \phi_\theta \right] \right\} R dx d\theta$$

$$\int_{-h/2}^{h/2} \int_0^{2\pi} \int_0^L \left\{ -D_x \left[\cos\left(\frac{\pi z}{h} \right) \frac{\partial \varphi}{\partial x} \right] - D_\theta \left[\cos\left(\frac{\pi z}{h} \right) \frac{\partial \varphi}{R \partial \theta} \right] \right. \qquad (10.28)$$

$$\left. - D_z \left[-\frac{\pi}{h} \sin\left(\frac{\pi z}{h} \right) \varphi - \frac{2V_0}{h} \right] \right\} R dx d\theta dz,$$

where the stress resultant–displacement relations can be written as:

$$\begin{Bmatrix} N_{xx} \\ N_{\theta\theta} \\ N_{x\theta} \end{Bmatrix} = \int_{-\frac{h}{2}}^{\frac{h}{2}} \begin{Bmatrix} \sigma_{xx} \\ \sigma_{\theta\theta} \\ \tau_{x\theta} \end{Bmatrix} dz , \qquad (10.29)$$

$$\begin{Bmatrix} Q_x \\ Q_\theta \end{Bmatrix} = k' \int_{-\frac{h}{2}}^{\frac{h}{2}} \begin{Bmatrix} \sigma_{xz} \\ \sigma_{\theta z} \end{Bmatrix} dz , \qquad (10.30)$$

$$\begin{Bmatrix} M_{xx} \\ M_{\theta\theta} \\ M_{x\theta} \end{Bmatrix} = \int_{-\frac{h}{2}}^{\frac{h}{2}} \begin{Bmatrix} \sigma_{xx} \\ \sigma_{\theta\theta} \\ \tau_{x\theta} \end{Bmatrix} z dz . \qquad (10.31)$$

In which k' is the shear correction coefficient. Substituting Eqs. (10.5)–(10.12) into Eqs. (1.29)–(10.31), the stress resultant–displacement relations can be obtained as:

$$N_{xx} = A_{110}\frac{\partial u}{\partial x} + A_{111}\frac{\partial \phi_x}{\partial x} + A_{120}\left(\frac{\partial v}{R\partial \theta} + \frac{w}{R}\right) + A_{121}\frac{\partial \phi_\theta}{R\partial \theta} + E_{31}\phi, \qquad (10.32)$$

$$N_{\theta\theta} = A_{120}\frac{\partial u}{\partial x} + A_{121}\frac{\partial \phi_x}{\partial x} + A_{220}\left(\frac{\partial v}{R\partial \theta} + \frac{w}{R}\right) + A_{221}\frac{\partial \phi_\theta}{R\partial \theta} + E_{32}\phi, \qquad (10.33)$$

$$Q_\theta = k'A_{44}\left[\frac{1}{R}\left(\frac{\partial w}{\partial \theta} - v\right) + \phi_\theta\right] + E_{15}\frac{\partial \phi}{R\partial \theta}, \qquad (10.34)$$

$$Q_x = k'A_{55}\left(\frac{\partial w}{\partial x} + \phi_x\right) + E_{24}\frac{\partial \phi}{\partial x}, \qquad (10.35)$$

$$N_{x\theta} = A_{660}\left(\frac{\partial u}{R\partial \theta} + \frac{\partial v}{\partial x}\right) + A_{661}\left(\frac{\partial \phi_x}{R\partial \theta} + \frac{\partial \phi_\theta}{\partial x}\right), \qquad (10.36)$$

$$M_{xx} = A_{111}\frac{\partial u}{\partial x} + A_{112}\frac{\partial \phi_x}{\partial x} + A_{121}\left(\frac{\partial v}{R\partial \theta} + \frac{w}{R}\right) + A_{122}\frac{\partial \phi_\theta}{R\partial \theta} + F_{31}\phi, \qquad (10.37)$$

$$M_{\theta\theta} = A_{121}\frac{\partial u}{\partial x} + A_{122}\frac{\partial \phi_x}{\partial x} + A_{221}\left(\frac{\partial v}{R\partial \theta} + \frac{w}{R}\right) + A_{222}\frac{\partial \phi_\theta}{R\partial \theta} + F_{32}\phi, \qquad (10.38)$$

$$M_{x\theta} = A_{661}\left(\frac{\partial u}{R\partial \theta} + \frac{\partial v}{\partial x}\right) + A_{662}\left(\frac{\partial \varphi_x}{R\partial \theta} + \frac{\partial \varphi_\theta}{\partial x}\right), \qquad (10.39)$$

where

$$A_{11k} = \int_{-h/2}^{h/2} C_{11}z^k dz, \qquad\qquad k = 0,1,2 \qquad (10.40)$$

$$A_{12k} = \int_{-h/2}^{h/2} C_{12}z^k dz, \qquad\qquad k = 0,1,2 \qquad (10.41)$$

$$A_{22k} = \int_{-h/2}^{h/2} C_{22}z^k dz, \qquad\qquad k = 0,1,2 \qquad (10.42)$$

$$A_{66k} = \int_{-h/2}^{h/2} C_{66}z^k dz, \qquad\qquad k = 0,1,2 \qquad (10.43)$$

$$A_{44} = \int_{-h/2}^{h/2} C_{44}dz, \qquad\qquad (10.44)$$

$$A_{55} = \int_{-h/2}^{h/2} C_{55} z^k dz,$$ (10.45)

$$(E_{31}, E_{32}) = \frac{\pi}{h} \int_{-h/2}^{h/2} (e_{31}, e_{32}) \sin\left(\frac{\pi z}{h}\right) dz,$$ (10.46)

$$(E_{24}, E_{15}) = -\int_{-h/2}^{h/2} (e_{24}, e_{15}) \cos\left(\frac{\pi z}{h}\right) dz,$$ (10.47)

$$(F_{31}, F_{32}) = \frac{\pi}{h} \int_{-h/2}^{h/2} (e_{31}, e_{32}) \sin\left(\frac{\pi z}{h}\right) z dz.$$ (10.48)

The kinetic energy of the system may be written as:

$$K = \frac{\rho}{2} \int \left[\left(\frac{\partial u}{\partial t} + z \frac{\partial \varphi_x}{\partial t}\right)^2 + \left(\frac{\partial v}{\partial t} + z \frac{\partial \varphi_\theta}{\partial t}\right)^2 + \left(\frac{\partial w}{\partial t}\right)^2 \right] dV.$$ (10.49)

Defining the moments of inertia as follows:

$$\begin{Bmatrix} I_0 \\ I_1 \\ I_2 \end{Bmatrix} = \int_{z^{(k-1)}}^{z^{(k)}} \begin{bmatrix} \rho \\ \rho z \\ \rho z^2 \end{bmatrix} dz,$$ (10.50)

the kinetic energy may be written as:

$$K = \frac{1}{2} \int \begin{pmatrix} I_0 \left[\left(\frac{\partial u}{\partial t}\right)^2 + \left(\frac{\partial v}{\partial t}\right)^2 + \left(\frac{\partial w}{\partial t}\right)^2 \right] + I_1 \left(2 \frac{\partial u}{\partial t} \frac{\partial \varphi_x}{\partial t} + 2 \frac{\partial v}{\partial t} \frac{\partial \varphi_\theta}{\partial t} \right) \\ + I_2 \left[\left(\frac{\partial \varphi_x}{\partial t}\right)^2 + \left(\frac{\partial \varphi_\theta}{\partial t}\right)^2 \right] \end{pmatrix} dA.$$ (10.51)

The external work due to the Pasternak medium can be written as [15]:

$$W_e = \int_0^{2\pi} \int_0^L \left(-K_w w + K_g \nabla^2 w \right) dA,$$ (10.52)

where K_w and K_g are Winkler's spring modulus and shear layer coefficients, respectively. The foundation stiffness K_w for a soft medium may be written by:

$$K_w = \frac{E_0}{4L(1-\nu_0^2)(2-c_1)^2} \left[5 - \left(2\gamma_1^2 + 6\gamma_1 + 5\right) \exp(-2\gamma_1) \right],$$ (10.53)

where

$$c_1 = (\gamma_1 + 2)\exp(-\gamma_1), \tag{10.54}$$

$$\gamma_1 = \frac{H_s}{L}, \tag{10.55}$$

$$E_0 = \frac{E_s}{(1 - \nu_s^2)}, \tag{10.56}$$

$$\nu_0 = \frac{\nu_s}{(1 - \nu_s)}, \tag{10.57}$$

where E_s, ν_s, and H_s are the Young's modulus, Poisson's ratio, and depth of the foundation, respectively. In this chapter, E_s is assumed to be temperature dependent, while ν_s is assumed to be constant. In addition, the in-plane forces may be written as:

$$W = -\frac{1}{2} \int \left[N_{xx}^f \left(\frac{\partial w}{\partial x}\right)^2 + \frac{N_{\theta\theta}^f}{R^2}\left(\frac{\partial w}{\partial y}\right)^2 \right] R dx d\theta, \tag{10.58}$$

where

$$N_{xx}^f = N_{xx}^M + N_{xx}^T + N_{xx}^E, \tag{10.59}$$

$$N_{\theta\theta}^f = N_{\theta\theta}^M + N_{\theta\theta}^T + N_{\theta\theta}^E, \tag{10.60}$$

where $N_{\theta\theta}^M = N_{xx}^M = 0$ are mechanical forces; $N_{\theta\theta}^E = 2V_0 e_{32}, N_{xx}^M = 2V_0 e_{31}$ are electrical forces; and $N_{\theta\theta}^T, N_{xx}^T$ are thermal forces, which may be written as:

$$\begin{Bmatrix} N_{xx}^T \\ N_{\theta\theta}^T \end{Bmatrix} = \int_{-h/2}^{h/2} \begin{Bmatrix} C_{11}(T,z)\alpha_{xx}(z) + C_{12}(T,z)\alpha_{\theta\theta}(z) \\ C_{12}(T,z)\alpha_{xx}(z) + C_{22}(T,z)\alpha_{\theta\theta}(z) \end{Bmatrix} \Delta T \, dz. \tag{10.61}$$

The governing equations can be derived by Hamilton's principal as follows:

$$\int_0^t (\delta U - \delta K - \delta W - \delta W_e) dt = 0. \tag{10.62}$$

Substituting Eqs. (10.28), (10.51), (10.53), and (10.58) into Eq. (10.62) yields the following governing equations:

$$\delta u : \frac{\partial N_{xx}}{\partial x} + \frac{\partial N_{x\theta}}{R\partial\theta} = I_0 \frac{\partial^2 u}{\partial t^2} + I_1 \frac{\partial^2 \varphi_x}{\partial t^2}, \tag{10.63}$$

$$\delta v : \frac{\partial N_{x\theta}}{\partial x} + \frac{\partial N_{\theta\theta}}{R\partial\theta} + \frac{Q_\theta}{R} = I_0 \frac{\partial^2 v}{\partial t^2} + I_1 \frac{\partial^2 \varphi_\theta}{\partial t^2}, \tag{10.64}$$

$$\delta w : \frac{\partial Q_x}{\partial x} + \frac{\partial Q_\theta}{R\partial\theta} - \frac{N_{\theta\theta}}{R} + N_{xx}^f \frac{\partial^2 w}{\partial x^2} + N_{\theta\theta}^f \frac{\partial^2 w}{R^2 \partial\theta^2} - K_w w + K_g \nabla^2 w = I_0 \frac{\partial^2 w}{\partial t^2}, \tag{10.65}$$

$$\delta\varphi_x : \frac{\partial M_{xx}}{\partial x} + \frac{\partial M_{x\theta}}{R\partial\theta} - Q_x = I_2 \frac{\partial^2 \varphi_x}{\partial t^2} + I_1 \frac{\partial^2 u}{\partial t^2}, \tag{10.66}$$

$$\delta\varphi_\theta : \frac{\partial M_{x\theta}}{\partial x} + \frac{\partial M_{\theta\theta}}{R\partial\theta} - Q_\theta = I_2 \frac{\partial^2 \varphi_\theta}{\partial t^2} + I_1 \frac{\partial^2 v}{\partial t^2}, \tag{10.67}$$

$$\delta\phi : \int_{-h/2}^{h/2} \left\{ \left[\cos\left(\frac{\pi z}{h}\right) \frac{\partial D_x}{\partial x} \right] + \left[\cos\left(\frac{\pi z}{h}\right) \frac{\partial D_\theta}{R\partial\theta} \right] + D_z \left[\frac{\pi}{h} \sin\left(\frac{\pi z}{h}\right) \right] \right\} dz = 0. \tag{10.68}$$

Substituting Eqs. (10.32) to (10.39) into Eqs. (10.63) to (10.68), the governing equations can be obtained as:

$$A_{110} \frac{\partial^2 u}{\partial x^2} + A_{111} \frac{\partial^2 \phi_x}{\partial x^2} + A_{120} \left(\frac{\partial^2 v}{R\partial\theta\partial x} + \frac{1}{R}\frac{\partial w}{\partial x} \right) + A_{121} \frac{\partial^2 \phi_\theta}{R\partial\theta\partial x} + E_{31} \frac{\partial\phi}{\partial x}$$

$$+ A_{120} \frac{\partial^2 u}{R\partial x\partial\theta} + A_{121} \frac{\partial^2 \phi_x}{R\partial x\partial\theta} + A_{220} \left(\frac{\partial^2 v}{R^2 \partial\theta^2} + \frac{1}{R^2}\frac{\partial w}{\partial\theta} \right) + A_{221} \frac{\partial^2 \phi_\theta}{R^2 \partial\theta^2} \tag{10.69}$$

$$+ E_{32} \frac{\partial\phi}{\partial\theta} = I_0 \frac{\partial^2 u}{\partial t^2} + I_1 \frac{\partial^2 \phi_x}{\partial t^2},$$

$$A_{120} \frac{\partial^2 u}{R\partial x\partial\theta} + A_{121} \frac{\partial^2 \phi_x}{R\partial x\partial\theta} + A_{220} \left(\frac{\partial^2 v}{R^2 \partial\theta^2} + \frac{1}{R^2}\frac{\partial w}{\partial\theta} \right) + A_{221} \frac{\partial^2 \phi_\theta}{R^2 \partial\theta^2}$$

$$+ \frac{E_{32}}{R} \frac{\partial\phi}{\partial\theta} + A_{660} \left(\frac{\partial^2 u}{R\partial\theta\partial x} + \frac{\partial^2 v}{\partial x^2} \right) + A_{661} \left(\frac{\partial^2 \phi_x}{R\partial\theta\partial x} + \frac{\partial^2 \phi_\theta}{\partial x^2} \right) \tag{10.70}$$

$$= I_0 \frac{\partial^2 v}{\partial t^2} + I_1 \frac{\partial^2 \phi_\theta}{\partial t^2},$$

$$k^{'} A_{44} \left[\frac{1}{R^2} \left(\frac{\partial w}{\partial\theta} - \frac{\partial v}{\partial\theta} \right) + \frac{1}{R}\frac{\partial\phi_\theta}{\partial\theta} \right] + E_{15} \frac{\partial\phi}{R^2 \partial\theta^2} + k^{'} A_{55} \left(\frac{\partial^2 w}{\partial x^2} + \frac{\partial\phi_x}{\partial x} \right)$$

$$+ E_{24} \frac{\partial^2 \phi}{\partial x^2} - A_{120} \frac{\partial u}{\partial x} - A_{121} \frac{\partial\phi_x}{\partial x} - A_{220} \left(\frac{\partial v}{R\partial\theta} + \frac{w}{R} \right) - A_{221} \frac{\partial\phi_\theta}{R\partial\theta}, \tag{10.71}$$

$$N_{xx}^f \frac{\partial^2 w}{\partial x^2} + N_{\theta\theta}^f \frac{\partial^2 w}{R^2 \partial\theta^2} - K_w w + K_g \nabla^2 w = I_0 \frac{\partial^2 w}{\partial t^2},$$

$$A_{111}\frac{\partial^2 u}{\partial x^2} + A_{112}\frac{\partial^2 \phi_x}{\partial x^2} + A_{121}\left(\frac{\partial^2 v}{R\partial\theta\partial x} + \frac{1}{R}\frac{\partial w}{\partial x}\right) + A_{122}\frac{\partial^2 \phi_\theta}{R\partial\theta} + F_{31}\frac{\partial\phi}{\partial x} +$$

$$A_{661}\left(\frac{\partial^2 u}{R^2\partial\theta^2} + \frac{\partial^2 v}{R\partial x\partial\theta}\right) + A_{662}\left(\frac{\partial^2 \phi_x}{R^2\partial\theta^2} + \frac{\partial^2 \phi_\theta}{R\partial x\partial\theta}\right) - \dot{k}A_{55}\left(\frac{\partial w}{\partial x} + \phi_x\right) \quad (10.72)$$

$$-E_{24}\frac{\partial\phi}{\partial x} = I_2\frac{\partial^2 \phi_x}{\partial t^2} + I_1\frac{\partial^2 u}{\partial t^2},$$

$$A_{121}\frac{\partial^2 u}{R\partial x\partial\theta} + A_{122}\frac{\partial^2 \phi_x}{R\partial x\partial\theta} + A_{221}\left(\frac{\partial^2 v}{R^2\partial\theta^2} + \frac{1}{R^2}\frac{\partial w}{\partial\theta}\right) + A_{222}\frac{\partial^2 \phi_\theta}{R^2\partial\theta^2}$$

$$+\frac{F_{32}}{R}\frac{\partial\phi}{\partial\theta} + A_{661}\left(\frac{\partial^2 u}{R\partial\theta\partial x} + \frac{\partial^2 v}{\partial x^2}\right) + A_{662}\left(\frac{\partial^2 \phi_x}{R\partial\theta\partial x} + \frac{\partial^2 \phi_\theta}{\partial x^2}\right) \quad (10.73)$$

$$-\dot{k}A_{44}\left[\frac{1}{R}\left(\frac{\partial w}{\partial\theta} - v\right) + \phi_\theta\right] - E_{15}\frac{\partial\phi}{R\partial\theta} = I_2\frac{\partial^2 \phi_\theta}{\partial t^2} + I_1\frac{\partial^2 v}{\partial t^2},$$

$$\delta\phi: \ -E_{15}\left(\frac{\partial\phi_x}{\partial x} + \frac{\partial^2 w}{\partial x^2}\right) + \Xi_{11}\left(\frac{\partial^2 \varphi}{R\partial x\partial\theta}\right) - E_{24}\left(\frac{1}{R^2}\left(\frac{\partial^2 w}{\partial\theta^2} - \frac{\partial v}{\partial\theta}\right) + \frac{\partial\phi_\theta}{R\partial\theta}\right)$$

$$+\Xi_{22}\left(\frac{\partial^2 \varphi}{R^2\partial\theta^2}\right) + E_{31}\frac{\partial u}{\partial x} + F_{31}\frac{\partial\phi_x}{\partial x} + \frac{E_{32}}{R}\left(w + \frac{\partial v}{\partial\theta}\right) + \frac{F_{32}}{R}\frac{\partial\phi_\theta}{\partial\theta} - \Xi_{33}\varphi = 0, \quad (10.74)$$

where

$$(\Xi_{11}, \Xi_{22}) = \int_{-h/2}^{h/2} (\in_{11}, \in_{22})\cos^2\left(\frac{\pi z}{h}\right) dz, \quad (10.75)$$

$$(\Xi_{33}) = \frac{\pi^2}{h^2}\int_{-h/2}^{h/2} (\in_{33})\sin^2\left(\frac{\pi z}{h}\right) dz. \quad (10.76)$$

In this chapter, three types of boundary conditions are used, which are:

- **Simple–Simple (SS)**

$$x = 0, L \Rightarrow u = v = w = \phi_\theta = M_x = 0, \quad (10.77)$$

- **Clamped–Clamped (CC)**

$$x = 0, L \Rightarrow u = v = w = \phi_x = \phi_\theta = 0, \quad (10.78)$$

- **Clamped–Simple (CS)**

$$x = 0 \Rightarrow u = v = w = \phi_x = \phi_\theta = 0,$$
$$x = L \Rightarrow u = v = w = \phi_x = M_x = 0. \tag{10.79}$$

Based on the DCM presented in Chapter 2, the motion equations may be written in matrix form as follows:

$$\left(\underbrace{\begin{bmatrix} K_{bb} & K_{bd} \\ K_{db} & K_{dd} \end{bmatrix}}_{[K]} + \underbrace{\begin{bmatrix} M_{bb} & M_{bd} \\ M_{db} & M_{dd} \end{bmatrix}}_{[M]} \omega^2 \right) \begin{bmatrix} Y_b \\ Y_d \end{bmatrix} = \begin{bmatrix} 0 \\ 0 \end{bmatrix}, \tag{10.80}$$

where $[K]$ and $[M]$ are stiffness and mass matrixes, respectively; Y is the displacement vector ($Y = (u, v, w, \varphi_x, \varphi_\theta, \varphi)$); and subscripts of b and d are related to boundary and domain points, respectively. Finally, for calculating the frequency of the system (ω), the eigenvalue problem can be used.

10.3 NUMERICAL RESULTS

In this section, the effects of different parameters on the frequency of the system are shown. For this purpose, PZT-5A is selected for the piezoelectric cylindrical shell with the following temperature-dependent thermal, mechanical, and electrical properties as follows [15, 21]:

$$E_{11} = E_{110} (1 + E_{111} \Delta T), \tag{10.81}$$

$$E_{22} = E_{220} (1 + E_{221} \Delta T), \tag{10.82}$$

$$E_{33} = E_{330} (1 + E_{331} \Delta T), \tag{10.83}$$

$$G_{12} = G_{120} (1 + G_{121} \Delta T), \tag{10.84}$$

$$G_{13} = G_{130} (1 + G_{131} \Delta T), \tag{10.85}$$

$$G_{23} = G_{230} (1 + G_{231} \Delta T), \tag{10.86}$$

$$\alpha_{xx} = \alpha_{110} (1 + \alpha_{111} \Delta T), \tag{10.87}$$

$$\alpha_{\theta\theta} = \alpha_{220} (1 + \alpha_{221} \Delta T), \tag{10.88}$$

$$e_{31} = d_{31} C_{11} + d_{32} C_{12}, \tag{10.89}$$

$$e_{32} = d_{31} C_{12} + d_{32} C_{22}, \tag{10.90}$$

$$e_{33} = d_{33}C_{33}, \tag{10.91}$$

$$e_{24} = d_{24}C_{44}, \tag{10.92}$$

$$e_{15} = d_{15}C_{55}, \tag{10.93}$$

where the constants are listed in Table 10.1.

Furthermore, the material properties of CNTs as reinforcement and the efficiency parameter η_j are shown in Tables 10.2 and 10.3, respectively. Poly dimethylsiloxane (PDMS) is selected for elastomeric medium, with $\nu_s = 0.48$ and $E_s = (3.22 - 0.0034T)\,GPa$, where $T = T_0 + \Delta T$ and $T_0 = 300\,K$ (room temperature) [15].

The convergence and accuracy of the DCM in evaluating the dimensionless frequency of the structure is shown in Figure 10.2 for different grid point numbers. The high convergence rate of the method is quite evident, and it is proven that the results may be converged with 113 grid points.

In order to validate the results of this chapter, ignoring the CNT volume percent (i.e., $V_{CNT} = 0$), elastic medium (i.e., $K_w = K_g = 0$), and temperature dependency of the shell (i.e., $T = 300\,K$), the vibration of a piezoelectric cylindrical shell is investigated.

TABLE 10.1
Elastic, Piezoelectric, and Thermal Constants of PZT-5

Elastic constants	Piezoelectric constants	Thermal coefficients
$E_{110} = E_{220} = 61\,GPa$	$d_{31} = d_{32} = -1.71\times10^{-10}\,m/V$	$\alpha_{120} = \alpha_{220} = 0.9\times10^{-6}\,1/K$
$E_{330} = 53.2\,GPa$	$d_{24} = d_{15} = 5.84\times10^{-10}\,m/V$	$\alpha_{111} = \alpha_{221} = 0.0005$
$G_{120} = G_{130} = G_{230} = 24.2\,GPa$	$d_{33} = 3.74\times10^{-10}\,m/V$	
$v_{12} = 0.35$		
$v_{23} = v_{13} = 0.38$		
$E_{111} = -0.0005$		
$E_{221} = E_{331} = G_{121} = G_{131} = G_{231} = -0.0002$		

TABLE 10.2
Temperature-Dependent Material Properties of (10, 10) SWCNTs (L = 9.26 nm, R = 0.68 nm, h = 0.067 nm, v_{12}^{CNT} = 0.175) [15]

	MD (Liew et al. 2014)		Mixture rule			
V_{CNT}	$E_{11}\,GPa$	$E_{22}\,GPa$	$E_{11}\,GPa$	η_1	$E_{22}\,GPa$	η_2
0.11	94.8	2.2	94.57	0.149	2.2	0.934
0.14	120.2	2.3	120.09	0.150	2.3	0.942
0.17	145.6	3.5	145.08	0.149	3.5	1.381

TABLE 10.3

Comparisons of Young's Moduli for *Poly* m-phenylenevinylene (PmPV)/CNT Composites Reinforced by (10,10)-Tube Under *T* = 300 K [15]

Temperature (K)	$E_{11}^{CNT}\,TPa$	$E_{22}^{CNT}\,TPa$	$G_{12}^{CNT}\,TPa$	$\alpha_{12}^{CNT}\,(10^{-6}\,/\,K)$	$\alpha_{22}^{CNT}\,(10^{-6}\,/\,K)$
300	5.6466	7.0800	1.9445	3.4584	5.1682
500	5.5308	6.9348	1.9643	4.5361	5.0189
700	5.4744	6.8641	1.9644	4.6677	4.8943

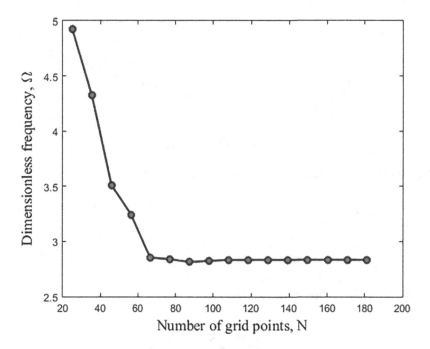

FIGURE 10.2 Convergence and accuracy of the DCM.

Based on the FSDT and Navier's method, the frequency of the structure is obtained and compared with the results of Sheng and Wang [19] in Figure 10.3. As can be seen, the present results are in good agreement with those reported by Sheng and Wang [19], indicating the validity of present work.

The effect of CNT distribution type in a piezoelectric cylindrical shell on the dimensionless frequency $\left(\Omega = \omega L \sqrt{\rho_m / E_{11}^m}\right)$ of the system versus external applied voltage is shown in Figure 10.4. The CNT uniform distribution and three types of FG patterns, namely, FGV, FGO, and FGX, are considered. It can be seen that the dimensionless frequency decreases with changing external applied voltage from negative to positive values. In the other words, the dimensionless frequency of the structure when applying a negative external voltage is higher than the dimensionless frequency of the

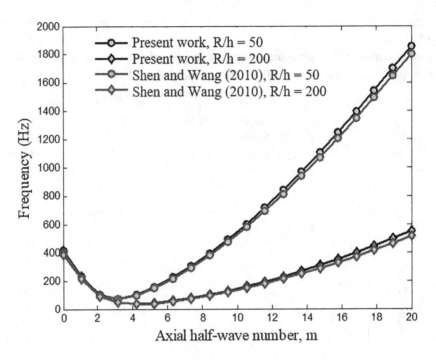

FIGURE 10.3 Comparison of the present work with that of Sheng and Wang [19].

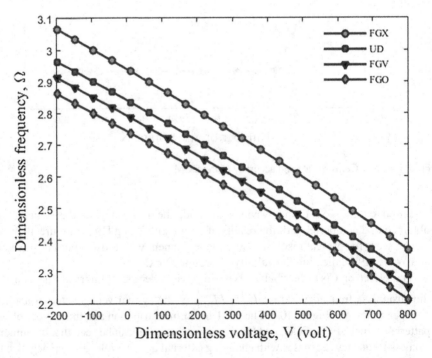

FIGURE 10.4 Dimension frequency versus external applied voltage for different CNT distribution types.

structure when applying a positive one. This is since the applied positive and negative voltages create the axial compressive and tensile forces in the structure, respectively.

With respect to the distribution types of CNTs in the shell, it can be concluded that the FGX pattern is the best choice compared to other cases. This is because, in the FGX mode, the frequency of the structure is maximum, which means the stiffness of the system is higher with respect to the other three patterns. Meanwhile, the frequency of the structure with uniform CNT distribution is higher than that of FGV and FGO models. However, it can be concluded that the CNTs distributed close to the top and bottom are more efficient than those distributed near the mid-plane.

The effect of the CNT volume fraction on the dimensionless frequency of the CNTRC shell with respect to externally applied voltage is illustrated in Figure 10.5. It can be found that applied negative voltage can increase the dimensionless frequency of the CNTRC shell. It is also observed that increasing the CNT volume fraction increases the dimensionless frequency of the structure. This is due to the fact that the increase in CNT volume fraction leads to a harder structure. Hence, applying nanotechnology in a shell is a new idea that can harden the system and consequently improve the vibration behavior of the structure.

The effect of the temperature-dependent elastic medium type on the dimensionless frequency of the CNTRC shell with respect to applied external voltage is illustrated in Figure 10.6. Three cases are considered: with no elastic medium $\left(K_w = 0 \text{ N/m}^3, K_g = 0 \text{ N/m}\right)$, Winkler medium $\left(K_w \neq 0 \text{ N/m}^3, K_g = 0 \text{ N/m}\right)$, and

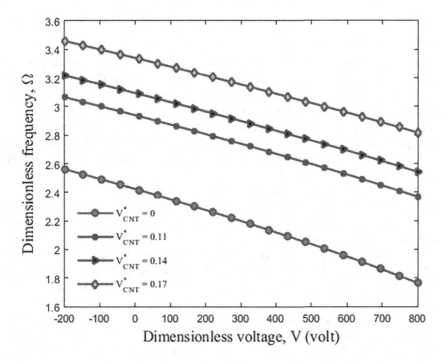

FIGURE 10.5 Dimension frequency versus external applied voltage for different CNT volume percents.

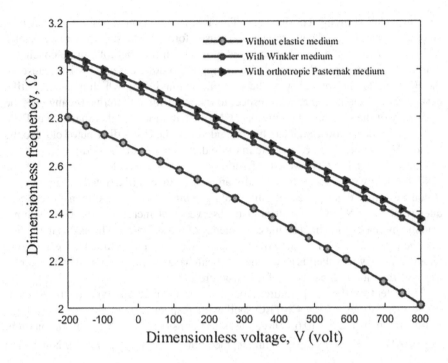

FIGURE 10.6 Dimension frequency versus external applied voltage for different elastic medium types.

Pasternak medium $\left(K_w \neq 0 \ \text{N/m}^3, K_g \neq 0 \ \text{N/m}\right)$. As can be seen, considering the elastic medium increases the dimensionless frequency of the CNTRC shell. This is due to the fact that considering the elastic medium leads to a stiffer structure. Furthermore, the effect of the Pasternak-type medium is greater than that of the Winkler-type medium on the dimensionless frequency of the CNTRC shell. This is perhaps due to the fact that the Winkler type is capable of describe just the normal load of the elastic medium, while the Pasternak type describes both transverse shear and normal loads of the elastic medium.

The effect of the temperature distribution type on the dimensionless frequency of the FG-CNT-reinforced piezoelectric cylindrical shell is demonstrated in Figure 10.7 versus applied external voltage. Here, three cases of uniform, linear, and harmonic distribution fields are considered. As can be seen, the dimensionless frequency of the structure subjected to harmonic temperature field is higher than that of the structures subjected to linear and uniform fields. In addition, the non-uniform temperature distributions lead to a higher frequency with respect to the structure under the uniform temperature field.

The effect of different boundary conditions on the dimensionless frequency of the system against external applied voltage is plotted in Figure 10.8. As can be seen, the structure with CC boundary condition has the maximum dimensionless frequency with respect to other cases. This is due to the fact that the CC boundary condition makes

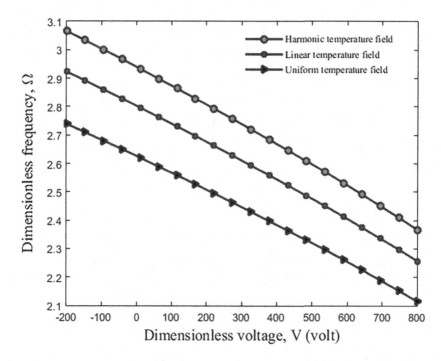

FIGURE 10.7 Dimension frequency versus external applied voltage for different temperature distribution types.

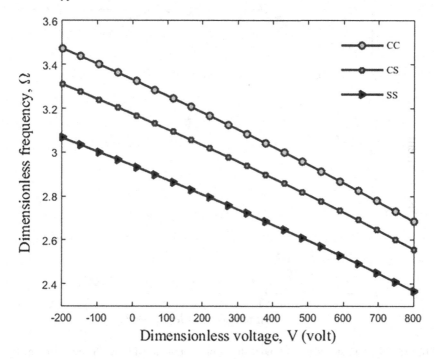

FIGURE 10.8 Dimension frequency versus external applied voltage for different boundary conditions.

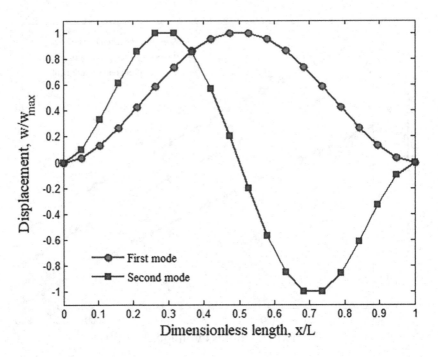

FIGURE 10.9 First- and second-mode shapes of the structure with the CC boundary condition.

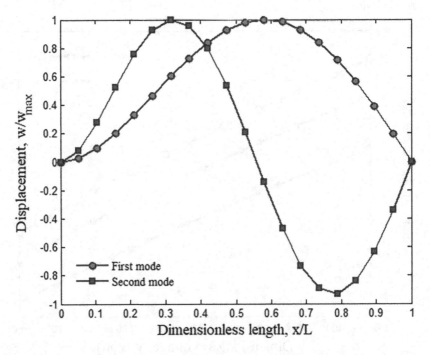

FIGURE 10.10 First- and second-mode shapes of the structure with the CS boundary condition.

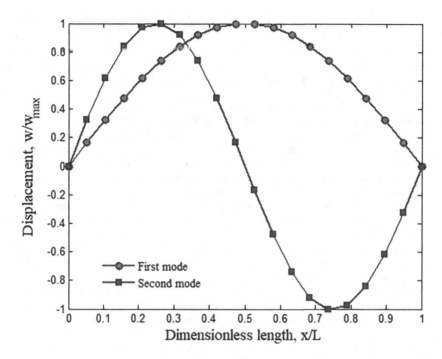

FIGURE 10.11 First- and second-mode shapes of the structure with the SS boundary condition.

the structure harder. Furthermore, the dimensionless frequency of the cylindrical shell with the assumed boundary conditions follows the following order: CC>CS>SS

The mode shapes of the structure for the first and second modes are shown in Figures 10.9–10.11, respectively, for CC, CS, and SS boundary conditions. It can be seen that the boundary conditions are satisfied at both ends of the cylindrical shell. Furthermore, the maximum deflection of the FG-CNT-reinforced cylindrical shell for the CC and SS boundary conditions is observed at the middle of the structure, while for the CS boundary condition, it is near the right side.

REFERENCES

1. Alibeigloo A, Nouri V. Static analysis of functionally graded cylindrical shell with piezoelectric layers using differential quadrature method. *Compos. Struct.* 2010;92:1775–1785.
2. Alibeigloo A. Static analysis of functionally graded carbon nanotube-reinforced composite plate embedded in piezoelectric layers by using theory of elasticity. *Compos. Struct.* 2013;95:612–622.
3. Bodaghi M, Shakeri M. An analytical approach for free vibration and transient response of functionally graded piezoelectric cylindrical panels subjected to impulsive loads. *Compos. Struct.* 2012;94:1721–1735.
4. Jinhua Y, Jie Y, Kitipornchai S. Dynamic stability of piezoelectric laminated cylindrical shells with delamination. *J. Intel. Mat. Sys. Struct.* 2013;24:1770–1781.
5. Nanda N, Nath Y. Active control of delaminated composite shells with piezoelectric sensor/actuator patches. *Struct. Eng. Mech.* 2012;42:211–228.

6. Lei ZX, Zhang LW, Liew KM, Yu JL. Dynamic stability analysis of carbonnanotube-reinforced functionally graded cylindrical panels using the element-free kp-Ritz method. *Compos. Struct.* 2014;113:328–338.

7. Lei ZX, Zhang LW, Liew KM. Free vibration analysis of laminated FG-CNT reinforced composite rectangular plates using the kp-Ritz method. *Compos. Struct.* 2015;127:245–259.

8. Lei ZX, Zhang LW, Liew KM. Elastodynamic analysis of carbon nanotube-reinforced functionally graded plates. *Int. J. Mech. Sci.* 2015;99:208–217.

9. Lei ZX, Zhang LW, Liew KM. Analysis of laminated CNT reinforced functionally graded plates using the element-free kp-Ritz method. *Compos. Part B: Eng.* 2016;84:211–221.

10. Liew KM, Lei ZX, Yu JL, Zhang LW. Postbuckling of carbon nanotube-reinforced functionally graded cylindrical panels under axial compression using a meshless approach. *Comput. Meth. Appl. Mech. Eng.* 2014;268:1–17.

11. Liu C, Ke LL, Wang YS, Yang J, Kitipornchai S. Thermo-electromechanical vibration of piezoelectric nanoplates based on the nonlocal theory. *Compos. Struct.* 2013; 106:167–174.

12. Rajesh Bhangale K, Ganesan N. Free vibration studies of simply supported non-homogeneous functionally graded magneto-electro-elastic finite cylindrical shells. *J. Sound Vib.* 2005;23:412–422.

13. Reddy JN. *Mechanics of laminated composite plates and shells: Theory and analysis.* 2nd edition. CRC Press, Boca Raton, USA, 2002.

14. Shen H. Postbuckling of FGM plates with piezoelectric actuators under thermo-electro-mechanical loadings. *Int. J. Solids Struct.* 2005;42:6101–6121.

15. Shen H. Nonlinear bending of functionally graded carbon nanotube-reinforced composite plates in thermal environments. *Compos. Struct.* 2009;91:9–19.

16. Shen H, Chen Z. Nonlocal beam model for nonlinear analysis of carbon nanotubes on elastomeric substrates. *Comput. Mat. Sci.* 2011;50:1022–1029.

17. Sheng GG, Wang X. Active control of functionally graded laminated cylindrical shells. *Compos. Struct.* 2009;90:448–457.

18. Sheng GG, Wang X. Studies on dynamic behavior of functionally graded cylindrical shells with PZT layers under moving loads. *J. Sound Vib.* 2009;323:772–789.

19. Sheng GG, Wang X. Thermoelastic vibration and buckling analysis of functionally graded piezoelectric cylindrical shells. *Appl. Math. Model.* 2010;34:2630–2643.

20. Tiersten HF. *Linear piezoelectric plate vibrations.* New York: Plenum Press, 1969.

21. Yaqoob Yasin M, Kapuria S. An efficient finite element with layerwise mechanics for smart piezoelectric composite and sandwich shallow shells. *Comput. Mech.* 2014;53:101–124.

22. Zhang LW, Zhu P, Liew KM. Thermal buckling of functionally graded plates using a local Kriging meshless method. *Compos. Struct.* 2014;108:472–492.

23. Zhang LW, Lei ZX, Liew KM, Yu JL. Large deflection geometrically nonlinear analysis of carbon nanotube-reinforced functionally graded cylindrical panels. *Comput. Meth. Appl. Mech. Eng.* 2014;273:1–18.

24. Zhang LW, Lei ZX, Liew KM, Yu JL. Static and dynamic of carbon nanotube reinforced functionally graded cylindrical panels. *Compos. Struct.* 2014;111:205–212.

25. Zhang LW, Liew KM. Large deflection analysis of FG-CNT reinforced composite skew plates resting on Pasternak foundations using an element-free approach. *Compos. Struct.* 2015;132:974–993.

26. Zhang LW, Liew KM. Geometrically nonlinear large deformation analysis of functionally graded carbon nanotube reinforced composite straight-sided quadrilateral plates. *Comput. Meth. Appl. Mech. Eng.* 2015;295:219–239.

27. Zhang LW, Li DM, Liew KM. An element-free computational framework for elastodynamic problems based on the IMLS-Ritz method. *Eng. Anal. Bound. Ele.* 2015;54:39–46.

28. Zhang LW, Huang D, Liew KM. An element-free IMLS-Ritz method for numerical solution of three-dimensional wave equations. *Comput. Meth. Appl. Mech. Eng.* 2015;297:116–139.
29. Zhang S, Schmidt R, Qin X. Active vibration control of piezoelectric bonded smart structures using PID algorithm. *Chin. J. Aero.* 2015;28:305–313.
30. Zhang LW, Song ZG, Liew KM. State-space Levy method for vibration analysis of FG-CNT composite plates subjected to in-plane loads based on higher-order shear deformation. *Compos. Struct.* 2015;134:989–1003.
31. Zhang LW, Lei ZX, Liew KM. Buckling analysis of FG-CNT reinforced composite thick skew plates using an element-free approach. *Compos. Part B: Eng.* 2015;75:36–46.
32. Zhang LW, Song ZG, Liew KM. Nonlinear bending analysis of FG-CNT reinforced composite thick plates resting on Pasternak foundations using the element-free IMLS-Ritz method. *Compos. Struct.* 2015;128:165–175.
33. Zhang LW, Cui WC, Liew KM. Vibration analysis of functionally graded carbon nanotube reinforced composite thick plates with elastically restrained edges. *Int. J. Mech. Sci.* 2015;103:9–21.
34. Zhang LW, Lei ZX, Liew KM. Free vibration analysis of functionally graded carbon nanotube-reinforced composite triangular plates using the FSDT and element-free IMLS-Ritz method. *Compos. Struct.* 2015;120:189–199.
35. Zhang LW, Liew KM. Element-free geometrically nonlinear analysis of quadrilateral functionally graded material plates with internal column supports. *Compos. Struct.* 2016;147:99–110.
36. Zhang LW, Liew KM, Reddy JN. Postbuckling behavior of bi-axially compressed arbitrarily straight-sided quadrilateral functionally graded material plates. *Comput. Meth. Appl. Mech. Eng.* 2016;300:593–610.
37. Zhang LW, Liew KM, Jiang Z. An element-free analysis of CNT-reinforced composite plates with column supports and elastically restrained edges under large deformation. *Compos. Part B: Eng.* 2016;95:18–28.
38. Zhang LW, Liew KM, Reddy JN. Postbuckling of carbon nanotube reinforced functionally graded plates with edges elastically restrained against translation and rotation under axial compression. *Comput. Meth. Appl. Mech. Eng.* 2016;298:1–28.
39. Zhang LW, Liew KM. Postbuckling analysis of axially compressed CNT reinforced functionally graded composite plates resting on Pasternak foundations using an element-free approach. *Compos. Struct.* 2016;138:40–51.
40. Zhang LW, Liew KM, Reddy JN. Postbuckling analysis of bi-axially compressed laminated nanocomposite plates using the first-order shear deformation theory. *Compos. Struct.* 2016;152:418–431.
41. Zhang LW, Song ZG, Liew KM. Computation of aerothermoelastic properties and active flutter control of CNT reinforced functionally graded composite panels in supersonic airflow. *Comput. Meth. Appl. Mech. Eng.* 2016;300:427–441.
42. Zhang LW, Song ZG, Liew KM. Optimal shape control of CNT reinforced functionally graded composite plates using piezoelectric patches. *Compos. Part B: Eng.* 2016;85:140–149.
43. Zhang LW, Xiao LN, Zou GL, Liew KM. Elastodynamic analysis of quadrilateral CNT-reinforced functionally graded composite plates using FSDT element-free method. *Compos. Struct.* 2016;148:144–154.
44. Zhang LW, Zhang Y, Zou GL, Liew KM. Free vibration analysis of triangular CNT-reinforced composite plates subjected to in-plane stresses using FSDT element-free method. *Compos. Struct.* 2016;149:247–260.
45. Zhu P, Zhang LW, Liew KM. Geometrically nonlinear thermomechanical analysis of moderately thick functionally graded plates using a local Petrov–Galerkin approach with moving Kriging interpolation. *Compos. Struct.* 2014;107:298–314.

11 Forced Vibration of Nanocomposite Shells

11.1 INTRODUCTION

CNTs have superior properties such as high tensile strength, high aspect ratio, high stiffness, and low density and can be used as the reinforcement phase for composite materials. However, in this chapter, the effect of CNTs on the forced vibration of a micro cylindrical shell is presented. In recent years, there has been significant research interest in investigating the mechanical behavior of various structural elements such as beams, plates, shells, and cylindrical structures. These studies are crucial due to the wide-ranging applications of these structures in engineering and technology. Functionally graded materials (FGMs) have emerged as promising candidates for such applications owing to their unique properties, which can be tailored to meet specific performance requirements. In this introduction, we will discuss several recent papers that delve into the size-dependent mechanical behavior, forced vibration response, and thermal stability of FGM structures using advanced analytical and numerical techniques [1].

Akbarov and Mehdiyev [2] explored the forced vibration behavior of an elastic system composed of a hollow cylinder and an encasing elastic medium under both perfect and imperfect contact conditions. Akbaş [3] performed an analysis on the forced vibration of functionally graded porous deep beams. Attia et al. [4] carried out a free vibration analysis of functionally graded plates with temperature-dependent properties using various four-variable refined plate theories. Belabed et al. [5] introduced an efficient higher-order shear and normal deformation theory for functionally graded material (FGM) plates. Beldjelili et al. [6] studied the hygro-thermo-mechanical bending of S-FGM plates resting on variable elastic foundations, applying a four-variable trigonometric plate theory. Belkorissat et al. [7] examined the vibration characteristics of functionally graded nanoplates using a novel nonlocal refined four-variable model. Bellifa et al. [8] and [9] conducted studies on the bending, free vibration analysis, and nonlinear postbuckling of nanobeams using various theoretical models. Bennoun et al. [10] proposed a new five-variable refined plate theory for analyzing the vibration of functionally graded sandwich plates. Bessaim et al. [11] presented a higher-order shear and normal deformation theory for the static and free vibration analysis of sandwich plates with functionally graded isotropic face sheets. Besseghier et al. [12] examined the free vibration of embedded nanosized FG plates using a new nonlocal trigonometric shear deformation theory. Bouafia et al. [13] proposed a nonlocal quasi-3D theory for analyzing the bending and free flexural vibration behaviors of functionally graded nanobeams. Bousahla et al. [14] investigated the thermal stability of plates with functionally graded coefficients of thermal expansion. Bouderba et al. [15] studied the thermomechanical bending response of

DOI: 10.1201/9781003510710-11

FGM thick plates on Winkler–Pasternak elastic foundations. Bouderba et al. [16] also analyzed the thermal stability of functionally graded sandwich plates using a simple shear deformation theory. Boukhari et al. [17] developed an efficient shear deformation theory for the wave propagation of functionally graded material plates. Bounouara et al. [18] proposed a nonlocal zeroth-order shear deformation theory for the free vibration of functionally graded nanoscale plates on elastic foundations. Bourada et al. [19] introduced a new simple shear and normal deformation theory for functionally graded beams. Bhushan et al. [20] performed simultaneous planar free and forced vibration analysis of an electrostatically actuated beam oscillator. Chen et al. [21] investigated the forced vibration of surface foundations on a multi-layered half-space. Chen et al. [22] also studied free and forced vibrations of shear deformable functionally graded porous beams. Chikh et al. [23] conducted thermal buckling analysis of cross-ply laminated plates using a simplified higher-order shear deformation theory. Dai et al. [24] examined the influence of surface effects on the nonlinear forced vibration of cantilevered nanobeams. Draiche et al. [25] proposed a refined theory incorporating stretching effects for the flexural analysis of laminated composite plates. Duc et al. [26] explored the vibration and nonlinear dynamic response of imperfect three-phase polymer nanocomposite panels on elastic foundations under hydrodynamic loads. Duc et al. [27] investigated the thermal and mechanical stability of functionally graded carbon nanotubes (FG-CNT)-reinforced composite truncated conical shells supported by elastic foundations. Duc et al. [28] also studied the static response and free vibration of functionally graded carbon nanotube-reinforced composite rectangular plates on Winkler–Pasternak elastic foundations. Duc et al. [29] proposed a novel approach for examining the nonlinear dynamic response and vibration of imperfect FG-CNT reinforced composite double curved shallow shells under blast loads and temperature variations. Duc et al. [30] analyzed the nonlinear dynamic response and vibration of nanocomposite multilayer organic solar cells. El-Haina et al. [31] presented a straightforward analytical method for the thermal buckling of thick functionally graded sandwich plates. Ghulghazaryan [32] examined the forced vibrations of orthotropic shells accounting for viscous resistance. Khetir et al. [33] developed a new nonlocal trigonometric shear deformation theory for thermal buckling analysis of embedded nanosized FG plates. Larbi Chaht et al. [34] carried out bending and buckling analyses of functionally graded material (FGM) nanoscale beams considering thickness stretching effects. Lei et al. [35] studied the bending and vibration of functionally graded sinusoidal microbeams using strain gradient elasticity theory. Li et al. [36] analyzed the nonlinear forced vibration and stability of axially moving viscoelastic sandwich beams. Liew et al. [37] investigated the postbuckling behavior of carbon nanotube-reinforced functionally graded cylindrical panels under axial compression using a meshless approach. Madani et al. [38] utilized the differential cubature method for vibration analysis of embedded FG-CNT-reinforced piezoelectric cylindrical shells under uniform and non-uniform temperature distributions. Mahi et al. [39] proposed a hyperbolic shear deformation theory for the bending and free vibration analysis of isotropic, functionally graded, sandwich, and laminated composite plates. Mehri et al. [40] conducted buckling and vibration analysis of pressurized CNT-reinforced functionally graded truncated conical shells under axial compression using the HDQ method. Menasria et al. [41] developed a new higher-order shear deformation theory for thermal stability analysis of FG sandwich plates. Meziane et al. [42] proposed a refined theory for the buckling and free vibration of exponentially graded sandwich plates under various boundary conditions. Mohamed et al. [43] conducted a numerical analysis of the nonlinear free and forced vibrations of buckled curved beams on nonlinear elastic foundations. Mouffoki et al. [44] investigated the vibration behavior of nonlocal advanced nanobeams in a hygrothermal environment using a new two-unknown trigonometric shear deformation beam theory. Orhan [45] analyzed the free and forced vibration of a cracked cantilever beam. Repetto et al. [46] studied the forced vibrations of a cantilever beam. Simsek [47] investigated the forced vibration of an embedded single-walled carbon nanotube subjected to a moving load using nonlocal Timoshenko beam theory. Simsek and Reddy [48] analyzed the bending and vibration of functionally graded microbeams using a new higher-order beam theory and the modified couple stress theory. Simsek and Kocatürk [49] performed a nonlinear dynamic analysis of an eccentrically prestressed damped beam under a concentrated moving harmonic load. Su et al. [50] studied the free and forced vibrations of nanowires on elastic substrates. Thanh et al. [51] examined the nonlinear dynamic response and vibration of functionally graded carbon nanotube-reinforced composite (FG-CNTRC) shear deformable plates with temperature-dependent material properties. Uymaz [52] analyzed the forced vibration of functionally graded beams using nonlocal elasticity theory. Van Thu and Duc [53] investigated the nonlinear dynamic response and vibration of imperfect three-phase laminated nanocomposite cylindrical panels on elastic foundations in a thermal environment. Virgin and Plaut [54] explored the impact of axial loads on the forced vibrations of beams. Zemri et al. [55] assessed the mechanical behavior of functionally graded nanoscale beams using a refined nonlocal shear deformation theory. Zidi et al. [56] conducted a bending analysis of FGM plates under hygro-thermo-mechanical loading using a four-variable refined plate theory. In this chapter, the nonlinear forced vibration of a micro cylindrical shell reinforced by FG-CNTs is studied based on the sinusoidal beam model. The size-dependent parameters are considered based on strain gradient theory. Based on the energy method and Hamilton's principal, the motion equations are derived. Applying the DQM and Newmark method, the frequency response of the structure is obtained. The effects of CNT volume percent and distribution type, boundary conditions, size effect, and length-to-thickness ratio are discussed in detail.

11.2 GOVERNING EQUATIONS

Figure 11.1 shows the geometry of the embedded micro cylindrical shell with radius R, length L, and thickness h. The structure is reinforced by FG-CNTs and is subjected to a harmonic load.

In this chapter, the sinusoidal shell theory presented in Chapter 1 is used. Based on the strain gradient theory (SGT), the potential strain energy of the structure can be expressed as [35]:

$$U = \frac{1}{2} \int_V \left(\sigma_{ij}\varepsilon_{ij} + P_i\gamma_i + \tau_{ijk}^{(1)}\eta_{ijk}^{(1)} + m_{ij}\chi_{ij} \right)dV, \tag{11.1}$$

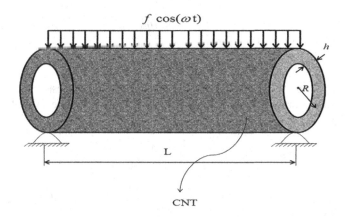

FIGURE 11.1 Schematic of the FG-CNT-reinforced cylindrical shell.

where ε_{ij}, γ_i, $\eta_{ijk}^{(1)}$, and χ_{ij} represent the strain, the dilatation gradient, the deviatoric stretch gradient, and the symmetric rotation gradient tensors, respectively, which are defined by:

$$\varepsilon_{ij} = \frac{1}{2}(\frac{\partial u_j}{\partial x_i} + \frac{\partial u_i}{\partial x_j}), \tag{11.2}$$

$$\gamma_i = \frac{\partial \varepsilon_{mm}}{\partial x_i}, \tag{11.3}$$

$$\eta_{ijk}^{(1)} = \frac{1}{3}\left(\frac{\partial \varepsilon_{jk}}{\partial x_i} + \frac{\partial \varepsilon_{ki}}{\partial x_j} + \frac{\partial \varepsilon_{ij}}{\partial x_k}\right) - \frac{1}{15}\left[\delta_{ij}\left(\frac{\partial \varepsilon_{mm}}{\partial x_k} + 2\frac{\partial \varepsilon_{mk}}{\partial x_m}\right) + \delta_{jk}\left(\frac{\partial \varepsilon_{mm}}{\partial x_i} + 2\frac{\partial \varepsilon_{mi}}{\partial x_m}\right)\right.$$
$$\left. + \delta_{ki}\left(\frac{\partial \varepsilon_{mm}}{\partial x_j} + 2\frac{\partial \varepsilon_{mj}}{\partial x_m}\right)\right], \tag{11.4}$$

$$\chi_{ij} = \frac{1}{2}\left(e_{ipq}\frac{\partial \varepsilon_{qj}}{\partial x_p} + e_{jpq}\frac{\partial \varepsilon_{qi}}{\partial x_p}\right), \tag{11.5}$$

where u_i, δ_{ij}, and e_{ijk} are the displacement vector, the knocker delta and the alternate tensor, respectively. The classical stress tensor σ_{ij} and the higher-order stresses p_i, $\tau_{ijk}^{(1)}$, and m_{ij} are given by:

$$\sigma_{ij} = E\delta_{ij}\varepsilon_{mm} + 2G\left(\varepsilon_{ij} - \frac{1}{3}\varepsilon_{mm}\delta_{ij}\right), \tag{11.6}$$

$$p_i = 2l_0^2 G\gamma_i, \tag{11.7}$$

$$\tau_{ijk}^{(1)} = 2l_1^2 G \eta_{ijk}^{(1)}, \tag{11.8}$$

$$m_{ij} = 2l_2^2 G \chi_{ij}, \tag{11.9}$$

where E and G are the bulk modulus and the shear modulus, respectively; (l_0, l_1, l_2) are independent material length scale parameters. The kinetic energy of the structure can be expressed as:

$$K = \frac{1}{2} \int_{-\frac{h}{2}}^{\frac{h}{2}} \int_0^{2\pi} \int_0^l \rho \left[\left(\frac{\partial u_x}{\partial t} \right)^2 + \left(\frac{\partial u_z}{\partial t} \right)^2 \right] dx d\theta dz , \tag{11.10}$$

where ρ denotes the density of the structure. The external work due to the harmonic transverse load can be given as:

$$W_t = -\int \left(f \cos(\omega t) \right) w dA, \tag{11.11}$$

where f and ω are amplitude and excitation frequency, respectively. However, based on Hamilton's principle, the motion equations can be written as:

$$\delta u: \left(\frac{4}{5} \frac{l_1^2 G \pi P_0}{h} \right) \frac{\partial^3 w}{\partial x^3} - \left(\frac{2h E P_0}{\pi} \right) \frac{\partial^3 w}{\partial x^3} + \left(-\frac{4}{5} \frac{l_1^2 G \pi P_0}{h} \right) \frac{\partial^2 \phi}{\partial x^2} + \left(\frac{8}{5} \frac{l_1^2 h G P_0}{\pi} + 4 \frac{l_0^2 h G P_0}{\pi} \right) \frac{\partial^5 w}{\partial x^5}$$

$$- \left(-2 \frac{h E P_0}{\pi} \right) \frac{\partial^2 \phi}{\partial x^2} + \left(-\frac{8}{5} \frac{l_1^2 h G P_0}{\pi} - 4 \frac{l_0^2 h G P_0}{\pi} \right) \frac{\partial^4 \phi}{\partial x^4} - (2EA) \frac{\partial^2 u}{\partial x^2} + \left(\frac{8}{5} l_1^2 GA + 2l_0^2 GA \right) \frac{\partial^4 u}{\partial x^4} + \rho A \frac{\partial^2 u}{\partial t^2} = 0,$$

$$\tag{11.12}$$

$$\delta w: \left(\frac{16}{15} \frac{l_1^2 G \pi^2 L}{h^2} + \frac{1}{2} \frac{l_2^2 G L \pi^2}{h^2} + 2GO \right) \frac{\partial \phi}{\partial x} - \left(\begin{array}{c} \frac{8}{5} l_1^2 GI + \frac{4}{5} \frac{l_1^2 h^2}{\pi^2} GI - \frac{8}{5} \frac{l_1^2 h}{\pi} GP_1 + 2l_0^2 GI \\ -4 \frac{l_0^2 h}{\pi} GP_1 + 2 \frac{l_0^2 h^2}{\pi^2} GL \end{array} \right) \frac{\partial^6 w}{\partial x^6}$$

$$- \left(-\frac{4}{5} l_1^2 GL + \frac{4}{5} \frac{l_1^2 G \pi P_1}{h} \right) \frac{\partial^3 \phi}{\partial x^3} - \left(\begin{array}{c} \frac{8}{5} \frac{l_1^2 h}{\pi} GP_0 \\ +4 \frac{l_0^2 h}{\pi} GP_0 \end{array} \right) \frac{\partial^5 u}{\partial x^5} + \left(\begin{array}{c} -\frac{64}{15} l_1^2 GO + \frac{32}{15} l_1^2 GT_0 - \frac{1}{2} l_2^2 GO + l_2^2 GT \\ +4 l_0^2 GT_0 - 4 l_0^2 GO - 2 \frac{h^2}{\pi^2} EL + 2 \frac{h}{\pi} EP \end{array} \right) \frac{\partial^3 \phi}{\partial x^3}$$

$$- \left(2 \frac{8}{15} \frac{l_1^2 G \pi^2 L}{h^2} + \frac{1}{4} \frac{l_2^2 G L \pi^2}{h^2} + GO \right) \frac{\partial^2 w}{\partial x^2} - \left(\begin{array}{c} -\frac{8}{5} \frac{l_1^2 h^2}{\pi^2} GL + \frac{8}{5} \frac{l_1^2 h}{\pi} GP_1 \\ -4 \frac{l_0^2 h^2}{\pi^2} GL + 4 \frac{l_0^2 h}{\pi} GP \end{array} \right) \frac{\partial^5 \phi}{\partial x^5} - \left(-\frac{4}{5} l_1^2 GL \right) \frac{\partial^3 \phi}{\partial x^3}$$

$$- \left(\frac{4}{5} \frac{l_1^2 G \pi P_0}{h} \right) \frac{\partial^3 u}{\partial x^3} + \left(2 \frac{h}{\pi} EP \right) \frac{\partial^3 u}{\partial x^3} - \left(\begin{array}{c} \frac{8}{5} l_1^2 GL \\ -\frac{4}{5} \frac{l_1^2 G \pi P_1}{h} \end{array} \right) \frac{\partial^4 w}{\partial x^4} + 2 \left(\begin{array}{c} -\frac{32}{15} l_1^2 GT + \frac{8}{15} l_1^2 GA + l_1^2 \frac{32}{15} GO - l_2^2 GT_0 \\ + l_2^2 \frac{1}{4} GO + l_2^2 GA - 4 l_0^2 GT_0 + 2 l_0^2 GO \\ + 2 l_0^2 GA + EI - 2 \frac{h}{\pi} EP_1 + \frac{h^2}{\pi^2} EL \end{array} \right) \frac{\partial^4 w}{\partial x^4}$$

$$- \frac{\rho h P_1}{\pi} \frac{\partial^3 \phi}{\partial x \partial t^2} - \frac{\rho h^2 L}{\pi^2} \frac{\partial^4 w}{\partial x^2 \partial t^2} + \frac{\rho h^2 L}{\pi^2} \frac{\partial^3 \phi}{\partial x \partial t^2} + \rho A \frac{\partial^2 w}{\partial t^2} = 0,$$

$$\tag{11.13}$$

$$\delta\phi: \quad -\left(\begin{array}{c}-l_1^2\dfrac{64}{15}GO+l_1^2\dfrac{32}{15}GT_0-l_2^2\dfrac{1}{2}GO+l_2^2GT_0\\+4l_0^4GT_0-4l_0^2GO-2\dfrac{h^3}{\pi^2}EL+2\dfrac{h}{\pi}EP_1\end{array}\right)\dfrac{\partial^3 w}{\partial x^3}+\left(\begin{array}{c}-\dfrac{8}{5}\dfrac{l_1^2 h^2}{\pi^2}GL+\dfrac{8}{5}\dfrac{l_1^2 h}{\pi}GP\\-4\dfrac{l_0^2 h^2}{\pi^2}GL+4\dfrac{l_0^2 h}{\pi}GP\end{array}\right)\dfrac{\partial^5 w}{\partial x^5}+\left(\begin{array}{c}-\dfrac{8}{5}\dfrac{l_1^2 h}{\pi}GP\\-4\dfrac{l_0^2 h}{\pi}GP_0\end{array}\right)\dfrac{\partial^4 u}{\partial x^4}$$

$$+2\left(\dfrac{4}{5}\dfrac{l_1^2 h^2}{\pi^2}GL+2\dfrac{l_0^2 h^2}{\pi^2}GL\right)\dfrac{\partial^4\phi}{\partial x^4}+\left(-\dfrac{16}{15}\dfrac{l_1^2 G\pi^2 L}{h^2}-\dfrac{1}{2}\dfrac{l_2^2 GL\pi^2}{h^2}-2GO\right)\dfrac{\partial w}{\partial x}+\left(-\dfrac{4}{5}l_1^2 GL\right)\dfrac{\partial^3 w}{\partial x^3}$$

$$+\left(-\dfrac{4}{5}\dfrac{l_1^2 G\pi P_0}{h}\right)\dfrac{\partial^2 u}{\partial x^2}+\left(-\dfrac{4}{5}l_1^2 GL+\dfrac{4}{5}\dfrac{l_1^2 G\pi P_1}{h}\right)\dfrac{\partial^3 w}{\partial x^3}+2\left(\dfrac{8}{15}\dfrac{l_1^2 G\pi^2 L}{h^2}+\dfrac{1}{4}\dfrac{l_2^2 GL\pi^2}{h^2}+GO\right)\phi+2\left(\dfrac{4}{5}l_1^2 GL\right)\dfrac{\partial^2\phi}{\partial x^2}$$

$$-2\left(l_1^2\dfrac{32}{15}GO+l_2^2\dfrac{1}{4}GO+2l_0^2 GO+\dfrac{h^2}{\pi^2}EL\right)\dfrac{\partial^2\phi}{\partial x^2}-\left(-2\dfrac{h}{\pi}EP_0\right)\dfrac{\partial^2 u}{\partial x^2}-\dfrac{\rho h P_0}{\pi}\dfrac{\partial^2 u}{\partial t^2}$$

$$+\dfrac{\rho h P_1}{\pi}\dfrac{\partial^3 w}{\partial x\partial t^2}-\dfrac{\rho h^2 L}{\pi^2}\dfrac{\partial^3 w}{\partial x\partial t^2}+\dfrac{\rho h^2 L}{\pi^2}\dfrac{\partial^2\phi}{\partial t^2}=0,$$

$$(11.14)$$

where the following integrals are defined:

$$\left(A_i, I, P_0, P_1, T_0, L, O\right)=\int_A \beta dA, \qquad (11.15)$$

where

$$\beta=\left(1, z^2, f^{(\sin)}, f^{(\cos)}, z f^{(\cos)}, \left(f^{(\sin)}\right)^2, \left(f^{(\cos)}\right)^2\right). \qquad (11.16)$$

In the present chapter, three different types of boundary conditions are considered:

- **Simple–Simple (SS)**

  @$x=0, L \Rightarrow u=w=M_x=0,$ (11.17)

- **Clamped–Clamped (CC)**

  @$x=0, L \Rightarrow u=w=\phi=0,$ (11.18)

- **Clamped–Simple (CS)**

  @$x=0 \Rightarrow u=w=\phi=0,$

  @$x=L \Rightarrow u=w=M_x=0.$ (11.19)

According to this theory, the effective Young's and shear moduli of the structure may be expressed as [37]:

$$E_{11}=\eta_1 V_{CNT}E_{r11}+(1-V_{CNT})E_m, \qquad (11.20)$$

$$\dfrac{\eta_2}{E_{22}}=\dfrac{V_{CNT}}{E_{r22}}+\dfrac{(1-V_{CNT})}{E_m}, \qquad (11.21)$$

$$\dfrac{\eta_3}{G_{12}}=\dfrac{V_{CNT}}{G_{r12}}+\dfrac{(1-V_{CNT})}{G_m}, \qquad (11.22)$$

where (E_{r1}, E_{r22}) and E_m are the Young's moduli of the CNTs and matrix, respectively; G_{r11} and G_m are the shear moduli of the CNTs and matrix, respectively; V_{CNT} and V_m show the volume fractions of the CNTs and matrix, respectively; and η_j $(j = 1, 2, 3)$ is the CNT efficiency parameter for considering the size-dependent material properties. Note that this parameter may be calculated using molecular dynamics (MD). However, the CNT distributions for the mentioned patterns obey the following relations:

$$UD: \quad V_{CNT} = V_{CNT}^*, \tag{11.23}$$

$$FGV: \quad V_{CNT}(z) = \left(1 + \frac{2z}{h}\right)V_{CNT}^*, \tag{11.24}$$

$$FGO: \quad V_{CNT}(z) = 2\left(1 - \frac{2|z|}{h}\right)V_{CNT}^*, \tag{11.25}$$

$$FGX: \quad V_{CNT}(z) = 2\left(\frac{2|z|}{h}\right)V_{CNT}^*, \tag{11.26}$$

where

$$V_{CNT}^* = \frac{w_{CNT}}{w_{CNT} + \left(\rho_{CNT}/\rho_m\right) - \left(\rho_{CNT}/\rho_m\right)w_{CNT}}, \tag{11.27}$$

where w_{CNT} is the mass fraction of the CNTs; ρ_m and ρ_{CNT} represent the densities of the matrix and CNTs, respectively; and ν_{r12} and ν_m are the Poisson's ratios of the CNT and matrix, respectively.

11.3 NUMERICAL RESULTS

The structure is made from poly methyl methacrylate (PMMA) with the constant Poisson's ratios of $\nu_m = 0.34$, temperature-dependent thermal coefficient of $\alpha_m = (1 + 0.0005\Delta T) \times 10^{-6}/K$, and temperature-dependent Young's modulus of $E_m = (3.52 - 0.0034T)\, GPa$, in which $T = T_0 + \Delta T$ and $T_0 = 300\ K$ (room temperature) [38].

The effect of the distribution type of CNTs on the frequency response of the structure is shown in Figure 11.2. The CNT uniform distribution and three types of FG patterns, namely, FGV, FGO, and FGX, are considered. It can be concluded that the FGX pattern is the best choice compared to other cases. This is because in the FGX mode, the frequency of the structure is maximum and the deflection is minimum with respect to the other cases. This means the stiffness of the system is higher with respect to other three patterns. Meanwhile, the frequency of the structure with CNT uniform distribution is higher than that in the FGV and FGO models. However, it can

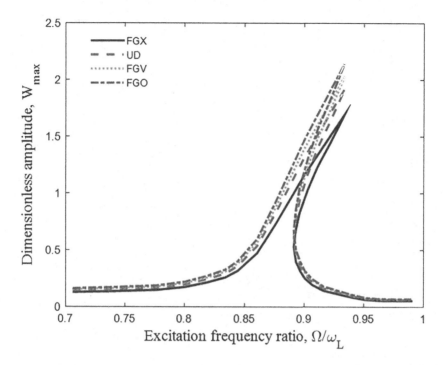

FIGURE 11.2 The effect of CNT distribution on the frequency response of the structure.

be concluded that the CNTs distributed close to the top and bottom are more efficient than those distributed near the mid-plane.

The effect of the CNT volume fraction on the frequency response of the cylindrical shell is illustrated in Figure 11.3. It is observed that increasing the CNT volume fraction increases the frequency and decreases the deflection of the structure. This is due to the fact that the increase in CNT volume fraction leads to a harder structure.

Figure 11.4 is plotted to study the effect of different theories of strain gradient, couple stress, and classical. As can be seen, the deflection of the strain gradient theory is lower than that for the couple stress theory, and the deflection of the couple stress is lower than that of the classical theory. This is because the strain gradient theory has three additional expressions consisting of the dilatation gradient tensor, the deviatoric stretch gradient tensor, and the rotation gradient tensor.

Figure 11.5 represents the effect of boundary conditions on the frequency response of the system. It can be seen that by considering the CC boundary condition, the maximum amplitude decreases and the frequency is increased. This is because the CC boundary condition leads to more bending rigidity.

The effect of the material length scale parameter on the frequency response of the structure is shown in Figure 11.6. As can be seen, by increasing the material length scale parameter, the amplitude of the system will be reduced.

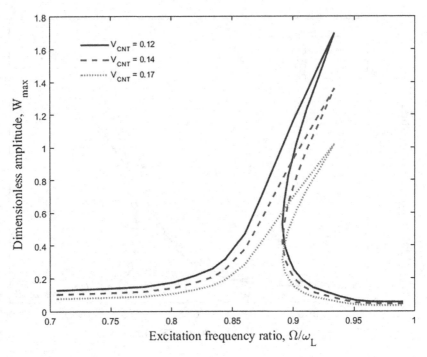

FIGURE 11.3 The effect of CNT volume percent on the frequency response of the structure.

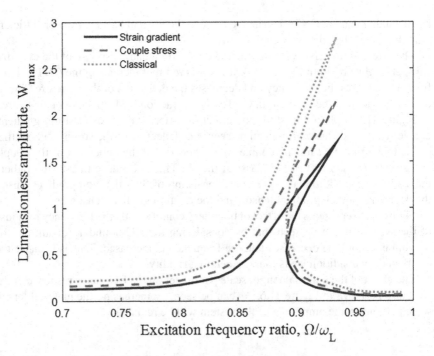

FIGURE 11.4 The effect of different theories on the frequency response of the structure.

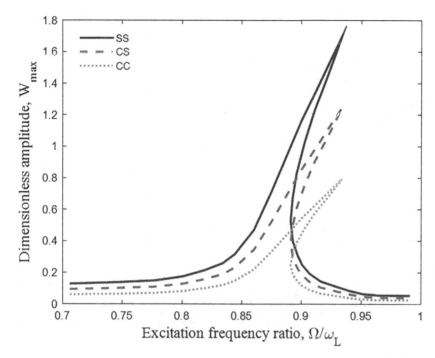

FIGURE 11.5 The effect of different boundary conditions on the frequency response of the structure.

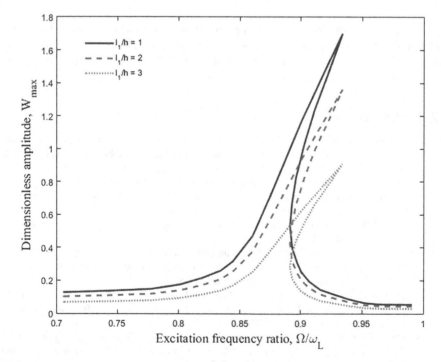

FIGURE 11.6 The effect of the material length scale parameter on the frequency response of the structure.

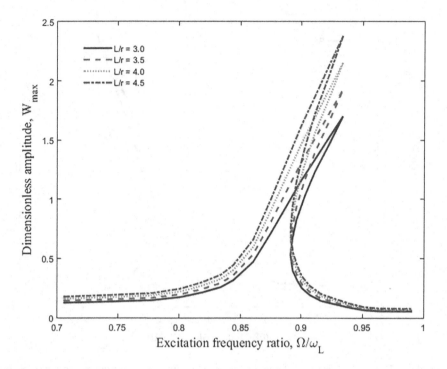

FIGURE 11.7 The effect of length-to-radius ratio on the frequency response of the structure.

The effects of the length-to-radius ratio of the cylindrical shell on the frequency response of the structure are presented in Figure 11.7. It is obvious that by increasing the length-to-radius ratio of the cylindrical shell, the structure becomes softer, and thus the deflection is increased.

REFERENCES

1. Ahouel M, Houari MSA, Adda Bedia EA, Tounsi A. Size-dependent mechanical behavior of functionally graded trigonometric shear deformable nanobeams including neutral surface position concept. *Steel Compos. Struct.* 2016;20(5):963–981.
2. Akbarov SD, Mehdiyev MA. Forced vibration of the elastic system consisting of the hollow cylinder and surrounding elastic medium under perfect and imperfect contact. *Struct. Eng. Mech.* 2017;62:113–123.
3. Akbaş ŞD. Forced vibration analysis of functionally graded porous deep beams. *Compos. Struct.* 2018;186:293–302.
4. Attia A, Tounsi A, Adda Bedia EA, Mahmoud SR. Free vibration analysis of functionally graded plates with temperature-dependent properties using various four variable refined plate theories. *Steel Compos. Struct.* 2015;18(1):187–212.
5. Belabed Z, Houari MSA, Tounsi A, Mahmoud SR, Bég OA. An efficient and simple higher order shear and normal deformation theory for functionally graded material (FGM) plates. *Compos. Part B.* 2014;60:274–283.
6. Beldjelili Y, Tounsi A, Mahmoud SR. Hygro-thermo-mechanical bending of S-FGM plates resting on variable elastic foundations using a four-variable trigonometric plate theory. *Smart. Struct. Syst.* 2016;18(4):755–786.

7. Belkorissat I, Houari MSA, Tounsi A, Hassan S. On vibration properties of functionally graded nano-plate using a new nonlocal refined four variable model. *Steel Compos. Struct.* 2015;18(4)·1063–1081.

8. Bellifa H, Benrahou KH, Hadji L, Houari MSA, Tounsi A. Bending and free vibration analysis of functionally graded plates using a simple shear deformation theory and the concept the neutral surface position. *J. Braz. Soc. Mech. Sci. Eng.* 2016;38(1):265–275.

9. Bellifa H, Benrahou KH, Bousahla AA, Tounsi A, Mahmoud SR. A nonlocal zeroth-order shear deformation theory for nonlinear postbuckling of nanobeams. *Struct. Eng. Mech.* 2017;62(6):695–702.

10. Bennoun M, Houari MSA, Tounsi A. A novel five variable refined plate theory for vibration analysis of functionally graded sandwich plates. *Mech. Advan. Mat. Struct.* 2016;23(4):423–431.

11. Bessaim A, Houari MSA, Tounsi A. A new higher-order shear and normal deformation theory for the static and free vibration analysis of sandwich plates with functionally graded isotropic face sheets. *J. Sandw. Struct. Mater.* 2013;15(6):671–703.

12. Besseghier A, Houari MSA, Tounsi A, Hassan S. Free vibration analysis of embedded nanosize FG plates using a new nonlocal trigonometric shear deformation theory. *Smart. Struct. Syst.* 2017;19(6):601–614.

13. Bouafia Kh, Kaci A, Houari MSA, Tounsi A. A nonlocal quasi-3D theory for bending and free flexural vibration behaviors of functionally graded nanobeams. *Smart. Struct. Syst.* 2017;19:115–126.

14. Bousahla AA, Benyoucef S, Tounsi A, Mahmoud SR. On thermal stability of plates with functionally graded coefficient of thermal expansion. *Struct. Eng. Mech.* 2016a;60(2):313–335.

15. Bouderba B, Houari MSA, Tounsi A. Thermomechanical bending response of FGM thick plates resting on Winkler–Pasternak elastic foundations. *Steel Compos. Struct.* 2013;14(1):85–104.

16. Bouderba B, Houari MSA, Tounsi A, Mahmoud SR. Thermal stability of functionally graded sandwich plates using a simple shear deformation theory. *Struct. Eng. Mech.* 2016b;58(3):397–422.

17. Boukhari A, Atmane HA, Tounsi A, Adda Bedia EA, Mahmoud SR. An efficient shear deformation theory for wave propagation of functionally graded material plates. *Struct. Eng. Mech.* 2016;57(5):837–859.

18. Bounouara F, Benrahou KH, Belkorissat I, Tounsi A. A nonlocal zeroth-order shear deformation theory for free vibration of functionally graded nanoscale plates resting on elastic foundation. *Steel Compos. Struct.* 2016;20(2):227–249.

19. Bourada M, Kaci A, Houari MSA, Tounsi A. A new simple shear and normal deformations theory for functionally graded beams. *Steel Compos. Struct.* 2015;18(2):409–423.

20. Bhushan A, Inamdar MM, Pawaskar DN. Simultaneous planar free and forced vibrations analysis of an electrostatically actuated beam oscillator. *Int. J. Mech. Sci.* 2014;82:90–99.

21. Chen L. Forced vibration of surface foundation on multi-layered half space. *Struct. Eng. Mech.* 2015;54:623–648.

22. Chen D, Yang J, Kitipornchai S. Free and forced vibrations of shear deformable functionally graded porous beams. *Int. J. Mech. Sci.* 2018;108–109:14–22.

23. Chikh A, Tounsi A, Hebali H, Mahmoud SR. Thermal buckling analysis of cross-ply laminated plates using a simplified HSDT. *Smart. Struct. Syst.* 2017;19(3):289–297.

24. Dai HL, Zhao DM, Zou JJ, Wang L. Surface effect on the nonlinear forced vibration of cantilevered nanobeams. *Physica E.* 2016;80:25–30.

25. Draiche K, Tounsi A, Mahmoud SR. A refined theory with stretching effect for the flexure analysis of laminated composite plates. *Geomech. Eng.* 2016;11:671–690.

26. Duc ND, Hadavinia H, Van Thu P, Quan TQ. Vibration and nonlinear dynamic response of imperfect three-phase polymer nanocomposite panel resting on elastic foundations under hydrodynamic loads. *Compos. Struct.* 2015;131:229–237.

27. Duc ND, Cong PH, Tuan ND, Tran P, Van Thanh N. Thermal and mechanical stability of functionally graded carbon nanotubes (FG CNT)-reinforced composite truncated conical shells surrounded by the elastic foundation. *Thin-Wall Struct.* 2017a;115:300–310.

28. Duc ND, Lee J, Nguyen-Thoi T, Thang PT. Static response and free vibration of functionally graded carbon nanotube-reinforced composite rectangular plates resting on Winkler–Pasternak elastic foundations. *Aerosp. Sci. Technol.* 2017b;68:391–402.

29. Duc ND, Tran QQ, Nguyen DK. New approach to investigate nonlinear dynamic response and vibration of imperfect functionally graded carbon nanotube reinforced composite double curved shallow shells subjected to blast load and temperature. *Aerosp. Sci. Technol.* 2017c;71:360–372.

30. Duc ND, Seung-Eock K, Quan TQ, Long DD, Anh VM. Nonlinear dynamic response and vibration of nanocomposite multilayer organic solar cell. *Compos. Struct.* 2018;184:1137–1144.

31. El-Haina F, Bakora A, Bousahla AA, Hassan S. A simple analytical approach for thermal buckling of thick functionally graded sandwich plates. *Struct. Eng. Mech.* 2017;63(5):585–595.

32. Ghulghazaryan LG. Forced vibrations of orthotropic shells when there is viscous resistance. *J. Appl. Math. Mech.* 2015;79:281–292.

33. Khetir H, Bouiadjra MB, Houari MSA, Tounsi A, Mahmoud SR. A new nonlocal trigonometric shear deformation theory for thermal buckling analysis of embedded nanosize FG plates. *Struct. Eng. Mech.* 2017;64(4):391–402.

34. Larbi Chaht F, Kaci A, Houari MSA, Hassan S. Bending and buckling analyses of functionally graded material (FGM) size-dependent nanoscale beams including the thickness stretching effect. *Steel Compos. Struct.* 2015;18(2):425–442.

35. Lei J, He Y, Zhang B, Gan Z, Zeng P. Bending and vibration of functionally graded sinusoidal microbeams based on the strain gradient elasticity theory. *Int. J. Eng. Sci.* 2013;72:36–52.

36. Li YH, Dong YH, Qin Y, Lv HW. Nonlinear forced vibration and stability of an axially moving viscoelastic sandwich beam. *Int. J. Mech. Sci.* 2018;138–139:131–145.

37. Liew KM, Lei ZX, Yu JL, Zhang LW. Postbuckling of carbon nanotube-reinforced functionally graded cylindrical panels under axial compression using a meshless approach. *Comput. Meth. Appl. Mech. Eng.* 2014;268:1–17.

38. Madani H, Hosseini H, Shokravi M. Differential cubature method for vibration analysis of embedded FG-CNT-reinForced piezoelectric cylindrical shells subjected to uniform and non-uniform temperature distributions. *Steel Compos. Struct.* 2016;22:889–913.

39. Mahi A, Bedia EAA, Tounsi A. A new hyperbolic shear deformation theory for bending and free vibration analysis of isotropic, functionally graded, sandwich and laminated composite plates. *Appl. Math. Model.* 2015;39:2489–2508.

40. Mehri M, Asadi H, Wang Q. Buckling and vibration analysis of a pressurized CNT reinforced functionally graded truncated conical shell under an axial compression using HDQ method. *Comput. Meth. Appl. Mech. Eng.* 2016;303:75–100.

41. Menasria A, Bouhadra A, Tounsi A, Hassan S. A new and simple HSDT for thermal stability analysis of FG sandwich plates. *Steel Compos. Struct.* 2017;25(2):157–175.

42. Meziane MAA, Abdelaziz HH, Tounsi AT. An efficient and simple refined theory for buckling and free vibration of exponentially graded sandwich plates under various boundary conditions. *J. Sandw. Struct. Mater.* 2014;16(3):293–318.

43. Mohamed N, Eltaher MA, Mohamed SA, Seddek LF. Numerical analysis of nonlinear free and forced vibrations of buckled curved beams resting on nonlinear elastic foundations. *Int. J. Non-Linear Mech.* 2018;101:157–173.

44. Mouffoki A, Adda Bedia EA, Houari MSA, Hassan S. Vibration analysis of nonlocal advanced nanobeams in hygro-thermal environment using a new two-unknown trigonometric shear deformation beam theory. *Smart. Struct. Syst.* 2017;20(3):369–383.

45. Orhan S. Analysis of free and forced vibration of a cracked cantilever beam. *NDT E. Int.* 2007;40:443–450.
46. Repetto CE, Roatta A, Welti RJ. Forced vibrations of a cantilever beam. *Europ. J. Phys.* 2012;33:345–366.
47. Simsek M. Forced vibration of an embedded single-walled carbon nanotube traversed by a moving load using nonlocal Timoshenko beam theory. *Steel Compos. Struct.* 2012;11:59–76.
48. Şimşek M, Reddy JN. Bending and vibration of functionally graded microbeams using a new higher order beam theory and the modified couple stress theory. *Int. J. Eng. Sci.* 2013;64:37–53.
49. Simsek M, Kocatu¨rk T. Nonlinear dynamic analysis of an eccentrically pre-stressed damped beam under a concentrated moving harmonic load. *J. Sound Vib.* 2009;320:235–253.
50. Su GY, Li YX, Li XY, Müller R. Free and forced vibrations of nanowires on elastic sub-strates. *Int. J. Mech. Sci.* 2018;138–139:62–73.
51. Thanh NV, Khoa ND, Tuan ND, Tran P, Duc ND. Nonlinear dynamic response and vibration of functionally graded carbon nanotube-reinforced composite (FG-CNTRC) shear deformable plates with temperature-dependent material properties. *J. Therm. Stres.* 2017;40:1254–1274.
52. Uymaz B. Forced vibration analysis of functionally graded beams using nonlocal elas-ticity. *Compos. Struct.* 2013;105:227–239.
53. Van Thu P, Duc ND. Non-linear dynamic response and vibration of an imperfect three-phase laminated nanocomposite cylindrical panel resting on elastic foundations in ther-mal environment. *Sci. Eng. Compos. Mat.* 2016;24(6):951–962.
54. Virgin LN, Plaut RH. Effect of axial load on forced vibrations of beams. *J. Sound Vib.* 1993;168:395–405.
55. Zemri A, Houari MSA, Bousahla AA, Tounsi A. A mechanical response of functionally graded nanoscale beam: An assessment of a refined nonlocal shear deformation theory beam theory. *Struct. Eng. Mech.* 2015;54(4):693–710.
56. Zidi M, Tounsi A, Bég OA. Bending analysis of FGM plates under hygro-thermo-mechanical loading using a four variable refined plate theory. *Aerosp. Sci. Tech.* 2014;34:24–34.

Appendix A

```
clc
clear all
syms P

Q=20;
QQ=linspace(1,20,Q);
for i=1:Q
gg=QQ(i);
n=1;
m=gg;
d=0.1;
a=10*d;
b=10*d;
Pi=pi;
c1=4/(3*(2*d)^2);
c2=3*c1;
KW=1e11;
KG=1e8;
alpha=1;

E1=132.38e9;
E2=10.76e9;
E3=10.76e9;
G12=3.61e9;
G13=5.65e9;
G23=5.65e9;
nu11=0.24;
nu23=0.24;
nu13=0.49;
nu31=nu13;
nu12=nu13;
nu21=nu12;
nu32=nu23;
rhot=1578;
rhom=1578;
rhob=1578;

Delta=1-nu12*nu21-nu23*nu32-nu31*nu13-2*nu12*nu32*nu13;
c11=E1*(1-nu23*nu32)/Delta;
c12=E1*(nu21+nu31*nu23)/Delta;
c13=E1*(nu31+nu21*nu32)/Delta;
c31=c13;
c22=E2*(1-nu13*nu31)/Delta;
c23=E2*(nu32+nu12*nu31)/Delta;
c32=c23;
c33=E3*(1-nu12*nu21)/Delta;
```

```
c44=G12;
c55=G13;
c66=G23;

t=0;
cc=cos(t);
s=sin(t);
Q11t=c11*cc^4+2*(c12+2*c44)*cc^2*s^2+c22*s^4;
Q12t=c12*(cc^4+s^4)+(c11+c22-4*c44)*cc^2*s^2;
Q13t=c13*cc^2+c23*s^2;
Q14t=(c11-c12-2*c44)*cc^3*s+(c12-c22+2*c44)*cc*s^3;
Q22t=c11*s^4+2*(c12+2*c44)*cc^2*s^2+c22*cc^4;
Q23t=c13*s^2+c23*cc^2;
Q24t=(c11-c12-2*c44)*cc*s^3+(c12-c22+2*c44)*cc^3*s;
Q23t=c33;
Q34t=(c31-c32)*cc*s;
Q44t=(c11-2*c12+c22-2*c44)*cc^2*s^2+c44*(cc^4+s^4);
Q55t=c55*cc^2+c66*s^2;
Q56t=(c55-c66)*cc*s;
Q66t=c66*s^2+c66*cc^2;

t=pi/2;
cc=cos(t);
s=sin(t);
Q11m=c11*cc^4+2*(c12+2*c44)*cc^2*s^2+c22*s^4;
Q12m=c12*(cc^4+s^4)+(c11+c22-4*c44)*cc^2*s^2;
Q13m=c13*cc^2+c23*s^2;
Q14m=(c11-c12-2*c44)*cc^3*s+(c12-c22+2*c44)*cc*s^3;
Q22m=c11*s^4+2*(c12+2*c44)*cc^2*s^2+c22*cc^4;
Q23m=c13*s^2+c23*cc^2;
Q24m=(c11-c12-2*c44)*cc*s^3+(c12-c22+2*c44)*cc^3*s;
Q23m=c33;
Q34m=(c31-c32)*cc*s;
Q44m=(c11-2*c12+c22-2*c44)*cc^2*s^2+c44*(cc^4+s^4);
Q55m=c55*cc^2+c66*s^2;
Q56m=(c55-c66)*cc*s;
Q66m=c66*s^2+c66*cc^2;

t=0;
cc=cos(t);
s=sin(t);
Q11b=c11*cc^4+2*(c12+2*c44)*cc^2*s^2+c22*s^4;
Q12b=c12*(cc^4+s^4)+(c11+c22-4*c44)*cc^2*s^2;
Q13b=c13*cc^2+c23*s^2;
Q14b=(c11-c12-2*c44)*cc^3*s+(c12-c22+2*c44)*cc*s^3;
Q22b=c11*s^4+2*(c12+2*c44)*cc^2*s^2+c22*cc^4;
Q23b=c13*s^2+c23*cc^2;
Q24b=(c11-c12-2*c44)*cc*s^3+(c12-c22+2*c44)*cc^3*s;
Q23b=c33;
Q34b=(c31-c32)*cc*s;
Q44b=(c11-2*c12+c22-2*c44)*cc^2*s^2+c44*(cc^4+s^4);
Q55b=c55*cc^2+c66*s^2;
```

```
Q56b=(c55-c66)*cc*s;
Q66b=c66*s^2+c66*cc^2;

K11=-(2/3)*Q11m*m^2*Pi^2*d/a^2-(2/3)*Q11t*m^2*Pi^2*d/a^2-
(2/3)*Q66m*n^2*Pi^2*d/b^2-(2/3)*Q11b*. . .
    m^2*Pi^2*d/a^2-(2/3)*Q66t*n^2*Pi^2*d/b^2-
(2/3)*Q66b*n^2*Pi^2*d/b^2;
K12=-(2/3)*Q66m*m*Pi^2*n*d/(a*b)-(2/3)*Q66t*m*Pi^2*n*d/(a*b)-
(2/3)*Q12t*m*Pi^2*n*d/(a*b)-(2/3)*Q66b*m*. . .
    Pi^2*n*d/(a*b)-(2/3)*Q12m*m*Pi^2*n*d/(a*b)-
(2/3)*Q12b*m*Pi^2*n*d/(a*b);
K13=-(20/81*(Q12b*c1*m*Pi^3*n^2/(a*b^2)+Q11b*c1*m^3*Pi^3/a^3))
*d^4+(40/81)*Q66t*c1*m*Pi^3*n^2*d^4/(a*b^2)+. . .
    (20/81*(Q12t*c1*m*Pi^3*n^2/(a*b^2)+Q11t*c1*m^3*Pi^3/
a^3))*d^4-(40/81)*Q66b*c1*m*Pi^3*n^2*d^4/(a*b^2);
K14=(20/81)*Q11t*c1*m^2*Pi^2*d^4/a^2-
(20/81)*Q66b*c1*n^2*Pi^2*d^4/b^2-(4/9)*Q66t*n^2*Pi^2*d^2/
b^2+(20/81)*Q66t*. . .
    c1*n^2*Pi^2*d^4/b^2+(4/9)*Q66b*n^2*Pi^2*d^2/b^2-
(4/9)*Q11t*m^2*Pi^2*d^2/a^2-(20/81)*Q11b*c1*m^2*Pi^2*d^4/
a^2+. . .
    (4/9)*Q11b*m^2*Pi^2*d^2/a^2;
K15=-(20/81)*Q12b*c1*m*Pi^2*n*d^4/
(a*b)+(4/9)*Q12b*m*Pi^2*n*d^2/(a*b)+(20/81)*Q12t*c1*m*Pi^2*n
*d^4/(a*b)-(20/81)*. . .
    Q66b*c1*m*Pi^2*n*d^4/(a*b)-(4/9)*Q66t*m*Pi^2*n*d^2/(a*b)+(
20/81)*Q66t*c1*m*Pi^2*n*d^4/(a*b)+(4/9)*Q66b*m*. . .
    Pi^2*n*d^2/(a*b)-(4/9)*Q12t*m*Pi^2*n*d^2/(a*b);

K21=-(2/3)*Q66m*m*Pi^2*n*d/(a*b)-(2/3)*Q66t*m*Pi^2*n*d/(a*b)-
(2/3)*Q12t*m*Pi^2*n*d/(a*b)-(2/3)*Q66b*m*Pi^2*n*. . .
    d/(a*b)-(2/3)*Q12m*m*Pi^2*n*d/(a*b)-(2/3)*Q12b*m*Pi^2*n*d/
(a*b);
K22=-(2/3)*Q22m*n^2*Pi^2*d/b^2-(2/3)*Q66m*m^2*Pi^2*d/a^2-
(2/3)*Q22b*n^2*Pi^2*d/b^2-(2/3)*Q66t*m^2*Pi^2*. . .
    d/a^2-(2/3)*Q66b*m^2*Pi^2*d/a^2-(2/3)*Q22t*n^2*Pi^2*d/b^2;
K23=(20/81*(Q22t*c1*n^3*Pi^3/b^3+Q12t*c1*m^2*Pi^3*n/
(a^2*b)))*d^4-(20/81*(Q22b*c1*n^3*Pi^3/b^3+Q12b*c1*. . .
    m^2*Pi^3*n/(a^2*b)))*d^4-(40/81)*Q66b*c1*m^2*Pi^3*n*d^4/
(a^2*b)+(40/81)*Q66t*c1*m^2*Pi^3*n*d^4/(a^2*b);
K24=-(20/81)*Q12b*c1*m*Pi^2*n*d^4/
(a*b)+(4/9)*Q12b*m*Pi^2*n*d^2/(a*b)+(20/81)*Q12t*c1*m*Pi^2*n
*d^4/(a*b)-(20/81)*. . .
    Q66b*c1*m*Pi^2*n*d^4/(a*b)-(4/9)*Q66t*m*Pi^2*n*d^2/(a*b)+(
20/81)*Q66t*c1*m*Pi^2*n*d^4/(a*b)+(4/9)*Q66b*. . .
    m*Pi^2*n*d^2/(a*b)-(4/9)*Q12t*m*Pi^2*n*d^2/(a*b);
K25=-(4/9)*Q22t*n^2*Pi^2*d^2/b^2+(4/9)*Q22b*n^2*Pi^2*d^2/
b^2-(20/81)*Q66b*c1*m^2*Pi^2*d^4/a^2+(20/81)*Q22t*. . .
    c1*n^2*Pi^2*d^4/b^2+(4/9)*Q66b*m^2*Pi^2*d^2/a^2-
(20/81)*Q22b*c1*n^2*Pi^2*d^4/b^2-(4/9)*Q66t*m^2*Pi^2*d^2/
a^. . .
    2+(20/81)*Q66t*c1*m^2*Pi^2*d^4/a^2;
```

```
K31=c1*((20/81)*Q11t*m^3*Pi^3*d^4/
a^3+(40/81)*Q66t*m*Pi^3*n^2*d^4/(a*b^2)-
(40/81)*Q66b*m*Pi^3*n^2*d^4/(a*b^2)-...
    (20/81)*Q12b*m*Pi^3*n^2*d^4/(a*b^2)-
(20/81)*Q11b*m^3*Pi^3*d^4/a^3+(20/81)*Q12t*m*Pi^3*n^2*d^4/
(a*b^2));
K32=c1*(-(20/81)*Q22b*n^3*Pi^3*d^4/
b^3+(40/81)*Q66t*m^2*Pi^3*n*d^4/(a^2*b)-
(40/81)*Q66b*m^2*Pi^3*n*d^4/(a^2*b)+...
    (20/81)*Q12t*m^2*Pi^3*n*d^4/(a^2*b)+(20/81)*Q22t*n^3*Pi^3
*d^4/b^3-(20/81)*Q12b*m^2*Pi^3*n*d^4/(a^2*b));
K33=(26/27)*Q55b*c1*m^2*Pi^2*d^3/
a^2+(2/27)*Q55m*c1*m^2*Pi^2*d^3/a^2+(26/27)*Q55t*c1*m^2*Pi^2
*d^3/a^2+(26/27)*Q44b*c1*n^2*Pi^2*d^3/b^2-
(2/3)*Q55b*m^2*Pi^2*d/a^2+(2/27)*Q44m*...
    c1*n^2*Pi^2*d^3/b^2-(2/3)*Q44m*n^2*Pi^2*d/b^2+(26/27)*Q44t
*c1*n^2*Pi^2*d^3/b^2-(2/3)*Q55m*m^2*Pi^2*d/a^2+c1*...
    ((2186/15309*(-Q22t*c1*n^4*Pi^4/b^4-Q12t*c1*m^2*Pi^4*n^2/
(a^2*b^2)))*d^7+(2186/15309*(-Q22b*c1*n^4*Pi^4/b^4-...
    Q12b*c1*m^2*Pi^4*n^2/(a^2*b^2)))*d^7-(8744/15309)*Q66b*c1*
m^2*Pi^4*n^2*d^7/(a^2*b^2)+(2186/15309*(-Q12b*c1*...
    m^2*Pi^4*n^2/(a^2*b^2)-Q11b*c1*m^4*Pi^4/a^4))*d^7-
(8744/15309)*Q66t*c1*m^2*Pi^4*n^2*d^7/
(a^2*b^2)+(2186/15309*...
    (-Q12t*c1*m^2*Pi^4*n^2/(a^2*b^2)-Q11t*c1*m^4*Pi^4/
a^4))*d^7+(2/15309*(-Q12m*c1*m^2*Pi^4*n^2/(a^2*b^2)-Q11m*...
    c1*m^4*Pi^4/a^4))*d^7+(2/15309*(-Q22m*c1*n^4*Pi^4/b^4-
Q12m*c1*m^2*Pi^4*n^2/(a^2*b^2)))*d^7-(8/15309)*Q66m*c1*...
    m^2*Pi^4*n^2*d^7/(a^2*b^2))-(2/3)*Q55t*m^2*Pi^2*d/a^2-
(2/3)*Q44b*n^2*Pi^2*d/b^2-c2*((242/405)*Q55b*c1*m^2*Pi^2*...
    d^5/a^2-(26/81)*Q55b*m^2*Pi^2*d^3/a^2+(2/405)*Q55m*c1*m^2*
Pi^2*d^5/a^2-(2/81)*Q55m*m^2*Pi^2*d^3/a^2+(242/405)*...
    Q55t*c1*m^2*Pi^2*d^5/a^2-(26/81)*Q55t*m^2*Pi^2*d^3/a^2)-
(2/3)*Q44t*n^2*Pi^2*d/b^2-c2*((242/405)*Q44b*c1*n^2*...
    Pi^2*d^5/b^2-(26/81)*Q44b*n^2*Pi^2*d^3/b^2+(2/405)*Q44m*c1
*n^2*Pi^2*d^5/b^2-(2/81)*Q44m*n^2*Pi^2*d^3/b^2+...
    (242/405)*Q44t*c1*n^2*Pi^2*d^5/b^2-
(26/81)*Q44t*n^2*Pi^2*d^3/b^2);
K34=-(2/3)*Q55b*m*Pi*d/
a+(26/27)*Q55b*c1*m*Pi*d^3/a-(2/3)*Q55t*m*Pi*d/a-
(2/3)*Q55m*m*Pi*d/a+(26/27)*Q55t*c1*m*...
    Pi*d^3/a+c1*(-(2186/15309)*Q11b*c1*m^3*Pi^3*d^7/a^3-
(2186/15309)*Q12b*c1*m*Pi^3*n^2*d^7/
(a*b^2)+(242/1215)*Q11b*...
    m^3*Pi^3*d^5/a^3+(4/1215)*Q66m*m*Pi^3*n^2*d^5/(a*b^2)-
(4372/15309)*Q66t*c1*m*Pi^3*n^2*d^7/
(a*b^2)+(242/1215)*Q11t*...
    m^3*Pi^3*d^5/a^3-(2186/15309)*Q11t*c1*m^3*Pi^3*d^7/a^3+(24
2/1215)*Q12b*m*Pi^3*n^2*d^5/(a*b^2)-(2/15309)*Q11m*c1*...
    m^3*Pi^3*d^7/a^3+(484/1215)*Q66b*m*Pi^3*n^2*d^5/(a*b^2)+(4
84/1215)*Q66t*m*Pi^3*n^2*d^5/(a*b^2)+(2/1215)*Q12m*...
    m*Pi^3*n^2*d^5/(a*b^2)+(2/1215)*Q11m*m^3*Pi^3*d^5/a^3-
```

```
(4372/15309)*Q66b*c1*m*Pi^3*n^2*d^7/(a*b^2)-(2186/15309)*. . .
    Q12t*c1*m*Pi^3*n^2*d^7/(a*b^2)-(2/15309)*Q12m*c1*m*Pi^3*n^
2*d^7/(a*b^2)+(242/1215)*Q12t*m*Pi^3*n^2*d^5/(a*. . .
    b^2)-(4/15309)*Q66m*c1*m*Pi^3*n^2*d^7/
(a*b^2))+(2/27)*Q55m*c1*m*Pi*d^3/a-c2*((242/405)*Q55b*c1*m*Pi*
d^5/a-(26/81)*. . .
    Q55b*m*Pi*d^3/a+(2/405)*Q55m*c1*m*Pi*d^5/a-
(2/81)*Q55m*m*Pi*d^3/
a+(242/405)*Q55t*c1*m*Pi*d^5/a-(26/81)*Q55t*. . .
    m*Pi*d^3/a)-KW-KG*((m*pi/a)^2+(n*pi/b)^2);
K35=(2/27)*Q44m*c1*n*Pi*d^3/b-(2/3)*Q44b*n*Pi*d/
b+(26/27)*Q44b*c1*n*Pi*d^3/b-(2/3)*Q44t*n*Pi*d/b+c1*(-
(2186/15309)*. . .
    Q12b*c1*m^2*Pi^3*n*d^7/(a^2*b)+(484/1215)*Q66t*m^2*Pi^3*n
*d^5/(a^2*b)-(2186/15309)*Q22b*c1*n^3*Pi^3*d^7/b^3+. . .
    (2/1215)*Q12m*m^2*Pi^3*n*d^5/(a^2*b)+(242/1215)*Q12b*m^2*P
i^3*n*d^5/(a^2*b)-(2186/15309)*Q22t*c1*n^3*Pi^3*. . .
    d^7/b^3+(4/1215)*Q66m*m^2*Pi^3*n*d^5/(a^2*b)+(242/1215)*Q2
2t*n^3*Pi^3*d^5/b^3-(4372/15309)*Q66t*c1*m^2*Pi^3*. . .
    n*d^7/(a^2*b)+(242/1215)*Q12t*m^2*Pi^3*n*d^5/(a^2*b)-
(2186/15309)*Q12t*c1*m^2*Pi^3*n*d^7/
(a^2*b)+(242/1215)*Q22b*. . .
    n^3*Pi^3*d^5/b^3-(2/15309)*Q12m*c1*m^2*Pi^3*n*d^7/(a^2*b)+
(484/1215)*Q66b*m^2*Pi^3*n*d^5/(a^2*b)+(2/1215)*Q22m*. . .
    n^3*Pi^3*d^5/b^3-(4372/15309)*Q66b*c1*m^2*Pi^3*n*d^7/
(a^2*b)-(2/15309)*Q22m*c1*n^3*Pi^3*d^7/b^3-
(4/15309)*Q66m*. . .
    c1*m^2*Pi^3*n*d^7/(a^2*b))-(2/3)*Q44m*n*Pi*d/
b+(26/27)*Q44t*c1*n*Pi*d^3/b-
c2*((242/405)*Q44b*c1*n*Pi*d^5/b-(26/81)*. . .
    Q44b*n*Pi*d^3/b+(2/405)*Q44m*c1*n*Pi*d^5/b-
(2/81)*Q44m*n*Pi*d^3/
b+(242/405)*Q44t*c1*n*Pi*d^5/b-(26/81)*Q44t*. . .
    n*Pi*d^3/b);

K41=(4/9)*Q66b*n^2*Pi^2*d^2/b^2-c2*((20/81)*Q66b*n^2*Pi^2*d^4/
b^2-(20/81)*Q66t*n^2*Pi^2*d^4/b^2)-c1*((20/81)*Q11b*. . .
    m^2*Pi^2*d^4/a^2-(20/81)*Q11t*m^2*Pi^2*d^4/
a^2)+(4/9)*Q11b*m^2*Pi^2*d^2/a^2-(4/9)*Q66t*n^2*Pi^2*d^2/
b^2-(4/9)*. . .
    Q11t*m^2*Pi^2*d^2/a^2;
K42=(4/9)*Q12b*m*Pi^2*n*d^2/(a*b)-(4/9)*Q12t*m*Pi^2*n*d^2/
(a*b)-(4/9)*Q66t*m*Pi^2*n*d^2/(a*b)-c2*((20/81)*Q66b*m*. . .
    Pi^2*n*d^4/(a*b)-(20/81)*Q66t*m*Pi^2*n*d^4/
(a*b))+(4/9)*Q66b*m*Pi^2*n*d^2/(a*b)-
c1*((20/81)*Q12b*m*Pi^2*n*d^4/(a*. . .
    b)-(20/81)*Q12t*m*Pi^2*n*d^4/(a*b));
K43=-c2*((4372/15309)*Q66b*c1*m*Pi^3*n^2*d^7/(a*b^2)+(4/15309)
*Q66m*c1*m*Pi^3*n^2*d^7/(a*b^2)+(4372/15309)*Q66t*. . .
    c1*m*Pi^3*n^2*d^7/(a*b^2))-(2/3)*Q55m*m*Pi*d/a-
```

```
(2/3)*Q55b*m*Pi*d/a+(4/1215)*Q66m*c1*m*Pi^3*n^2*d^5/
(a*b^2)+. . .
    (26/27)*Q55b*c1*m*Pi*d^3/a+(484/1215)*Q66b*c1*m*Pi^3*n^2
*d^5/(a*b^2)-c1*((2186/15309*(Q12b*c1*m*Pi^3*n^2/(a*. . .
    b^2)+Q11b*c1*m^3*Pi^3/a^3))*d^7+(2/15309*(Q12m*c1*m*Pi^3
*n^2/(a*b^2)+Q11m*c1*m^3*Pi^3/a^3))*d^7+(2186/15309*. . .
    (Q12t*c1*m*Pi^3*n^2/(a*b^2)+Q11t*c1*m^3*Pi^3/
a^3))*d^7)+c2*(-(242/405)*Q55b*c1*m*Pi*d^5/
a+(26/81)*Q55b*m*Pi*. . .
    d^3/a-(2/405)*Q55m*c1*m*Pi*d^5/a+(2/81)*Q55m*m*Pi*d^3/a-
(242/405)*Q55t*c1*m*Pi*d^5/a+(26/81)*Q55t*m*Pi*d^3/a)+. . .
    (2/27)*Q55m*c1*m*Pi*d^3/a-(2/3)*Q55t*m*Pi*d/a+(484/1215)*Q
66t*c1*m*Pi^3*n^2*d^5/(a*b^2)+(2/1215*(Q12m*c1*m*. . .
    Pi^3*n^2/(a*b^2)+Q11m*c1*m^3*Pi^3/a^3))*d^5+(26/27)*Q55t*c
1*m*Pi*d^3/a+(242/1215*(Q12b*c1*m*Pi^3*n^2/(a*b^2)+. . .
    Q11b*c1*m^3*Pi^3/a^3))*d^5+(242/1215*(Q12t*c1*m*Pi^3*n^2/
(a*b^2)+Q11t*c1*m^3*Pi^3/a^3))*d^5;
K44=-(26/81)*Q66b*n^2*Pi^2*d^3/b^2-c2*((2186/15309)*Q66b*c1*n^
2*Pi^2*d^7/b^2-(242/1215)*Q66t*n^2*Pi^2*d^5/b^2-. . .
    (242/1215)*Q66b*n^2*Pi^2*d^5/b^2+(2186/15309)*Q66t*c1*n^2*
Pi^2*d^7/b^2-(2/1215)*Q66m*n^2*Pi^2*d^5/b^2+(2/15309)*. . .
    Q66m*c1*n^2*Pi^2*d^7/b^2)-(2/81)*Q11m*m^2*Pi^2*d^3/a^2+(24
2/1215)*Q11b*c1*m^2*Pi^2*d^5/a^2+(2/1215)*Q66m*. . .
    c1*n^2*Pi^2*d^5/b^2+(26/27)*Q55b*c1*d^3+(2/1215)*Q11m*c1*m
^2*Pi^2*d^5/a^2+(242/1215)*Q66t*c1*n^2*Pi^2*d^5/b^2+. . .
    (242/1215)*Q66b*c1*n^2*Pi^2*d^5/b^2-(2/3)*Q55b*d-
(26/81)*Q11b*m^2*Pi^2*d^3/a^2-(2/3)*Q55m*d-
(2/81)*Q66m*n^2*. . .
    Pi^2*d^3/b^2+(2/27)*Q55m*c1*d^3-c1*((2186/15309)*Q11b*c1*m
^2*Pi^2*d^7/a^2-(242/1215)*Q11t*m^2*Pi^2*d^5/a^2-. . .
    (242/1215)*Q11b*m^2*Pi^2*d^5/a^2+(2186/15309)*Q11t*c1*m^2*
Pi^2*d^7/a^2-(2/1215)*Q11m*m^2*Pi^2*d^5/a^2+(2/15309)*. . .
    Q11m*c1*m^2*Pi^2*d^7/a^2)-(26/81)*Q66t*n^2*Pi^2*d^3/b^2+(2
42/1215)*Q11t*c1*m^2*Pi^2*d^5/a^2+(26/27)*Q55t*c1*. . .
    d^3-(2/3)*Q55t*d-(26/81)*Q11t*m^2*Pi^2*d^3/a^2+c2*(-
(242/405)*Q55b*c1*d^5+(26/81)*Q55b*d^3-(2/405)*Q55m*c1*. . .
    d^5+(2/81)*Q55m*d^3-(242/405)*Q55t*c1*d^5+(26/81)*Q55t
*d^3);
K45=-(26/81)*Q66b*m*Pi^2*n*d^3/(a*b)-(2/81)*Q12m*m*Pi^2*n*d^3/
(a*b)-(26/81)*Q66t*m*Pi^2*n*d^3/(a*b)+(242/1215)*. . .
    Q66t*c1*m*Pi^2*n*d^5/(a*b)-c2*((2186/15309)*Q66b*c1*m*Pi^2
*n*d^7/(a*b)-(242/1215)*Q66t*m*Pi^2*n*d^5/(a*b)-
(242/1215)*. . .
    Q66b*m*Pi^2*n*d^5/(a*b)+(2186/15309)*Q66t*c1*m*Pi^2*n*d^7/
(a*b)-(2/1215)*Q66m*m*Pi^2*n*d^5/(a*b)+(2/15309)*Q66m*. . .
    c1*m*Pi^2*n*d^7/(a*b))+(242/1215)*Q66b*c1*m*Pi^2*n*d^5/(a*
b)+(242/1215)*Q12b*c1*m*Pi^2*n*d^5/(a*b)-(26/81)*Q12b*m*. . .
    Pi^2*n*d^3/(a*b)+(2/1215)*Q66m*c1*m*Pi^2*n*d^5/(a*b)+(2/12
15)*Q12m*c1*m*Pi^2*n*d^5/(a*b)-c1*((2186/15309)*Q12b*. . .
    c1*m*Pi^2*n*d^7/(a*b))+(242/1215)*Q12t*m*Pi^2*n*d^5/(a*b)-
```

```
(242/1215)*Q12b*m*Pi^2*n*d^5/(a*b)+(2186/15309)*Q12t*c1*. . .
    m*Pi^2*n*d^7/(a*b)-(2/1215)*Q12m*m*Pi^2*n*d^5/(a*b)+(2/153
09)*Q12m*c1*m*Pi^2*n*d^7/(a*b))-(2/81)*Q66m*m*Pi^2*n*. . .
    d^3/(a*b)+(242/1215)*Q12t*c1*m*Pi^2*n*d^5/(a*b)-
(26/81)*Q12t*m*Pi^2*n*d^3/(a*b);

K51=-c2*((20/81)*Q12b*m*Pi^2*n*d^4/(a*b)-
(20/81)*Q12t*m*Pi^2*n*d^4/(a*b))-(4/9)*Q12t*m*Pi^2*n*d^2/
(a*b)-(4/9)*Q66t*. . .
    m*Pi^2*n*d^2/(a*b)+(4/9)*Q12b*m*Pi^2*n*d^2/
(a*b)+(4/9)*Q66b*m*Pi^2*n*d^2/(a*b)-
c1*((20/81)*Q66b*m*Pi^2*n*d^4/(a*. . .
    b)-(20/81)*Q66t*m*Pi^2*n*d^4/(a*b));
K52=-c2*((20/81)*Q22b*n^2*Pi^2*d^4/b^2-
(20/81)*Q22t*n^2*Pi^2*d^4/b^2)+(4/9)*Q66b*m^2*Pi^2*d^2/a^2-
c1*((20/81)*Q66b*. . .
    m^2*Pi^2*d^4/a^2-(20/81)*Q66t*m^2*Pi^2*d^4/
a^2)+(4/9)*Q22b*n^2*Pi^2*d^2/b^2-(4/9)*Q22t*n^2*Pi^2*d^2/
b^2-(4/9)*. . .
    Q66t*m^2*Pi^2*d^2/a^2;
K53=(2/1215*(Q22m*c1*n^3*Pi^3/b^3+Q12m*c1*m^2*Pi^3*n/
(a^2*b)))*d^5+c2*(-(242/405)*Q44b*c1*n*Pi*d^5/b+(26/81)*. . .
    Q44b*n*Pi*d^3/b-(2/405)*Q44m*c1*n*Pi*d^5/
b+(2/81)*Q44m*n*Pi*d^3/b-(242/405)*Q44t*c1*n*Pi*d^5/
b+(26/81)*Q44t*. . .
    n*Pi*d^3/b)+(242/1215*(Q22t*c1*n^3*Pi^3/
b^3+Q12t*c1*m^2*Pi^3*n/
(a^2*b)))*d^5+(26/27)*Q44b*c1*n*Pi*d^3/b-c2*. . .
    ((2186/15309*(Q22b*c1*n^3*Pi^3/b^3+Q12b*c1*m^2*Pi^3*n/
(a^2*b)))*d^7+(2/15309*(Q22m*c1*n^3*Pi^3/b^3+Q12m*. . .
    c1*m^2*Pi^3*n/(a^2*b)))*d^7+(2186/15309*(Q22t*c1*n^3*Pi^3/
b^3+Q12t*c1*m^2*Pi^3*n/(a^2*b)))*d^7)+(242/1215*(Q22b*. . .
    c1*n^3*Pi^3/b^3+Q12b*c1*m^2*Pi^3*n/(a^2*b)))*d^5+(4/1215)*
Q66m*c1*m^2*Pi^3*n*d^5/(a^2*b)-c1*((4372/15309)*Q66b*. . .
    c1*m^2*Pi^3*n*d^7/(a^2*b)+(4/15309)*Q66m*c1*m^2*Pi^3*n
*d^7/(a^2*b)+(4372/15309)*Q66t*c1*m^2*Pi^3*n*d^7/
(a^2*b))+. . .
    (484/1215)*Q66t*c1*m^2*Pi^3*n*d^5/(a^2*b)+(484/1215)*Q66b*
c1*m^2*Pi^3*n*d^5/(a^2*b)-(2/3)*Q44t*n*Pi*d/b-(2/3)*Q44b*. . .
    n*Pi*d/b+(26/27)*Q44t*c1*n*Pi*d^3/b-(2/3)*Q44m*n*Pi*d/
b+(2/27)*Q44m*c1*n*Pi*d^3/b;
K54=-c1*((2186/15309)*Q66b*c1*m*Pi^2*n*d^7/(a*b)-
(242/1215)*Q66t*m*Pi^2*n*d^5/(a*b)-
(242/1215)*Q66b*m*Pi^2*n*d^5/(a*b)+. . .
    (2186/15309)*Q66t*c1*m*Pi^2*n*d^7/(a*b)-
(2/1215)*Q66m*m*Pi^2*n*d^5/(a*b)+(2/15309)*Q66m*c1*m*Pi^2*n
*d^7/(a*b))+. . .
    (242/1215)*Q66b*c1*m*Pi^2*n*d^5/(a*b)+(242/1215)*Q12t*c1*m
*Pi^2*n*d^5/(a*b)+(242/1215)*Q12b*c1*m*Pi^2*n*d^5/(a*b)+. . .
    (2/1215)*Q12m*c1*m*Pi^2*n*d^5/(a*b)-
```

```
(26/81)*Q12t*m*Pi^2*n*d^3/(a*b)-(26/81)*Q12b*m*Pi^2*n*d^3/
(a*b)+(242/1215)*. . .
    Q66t*c1*m*Pi^2*n*d^5/(a*b)-(2/81)*Q12m*m*Pi^2*n*d^3/
(a*b)-(2/81)*Q66m*m*Pi^2*n*d^3/(a*b)-(26/81)*Q66t*m*Pi^2*. . .
    n*d^3/(a*b)+(2/1215)*Q66m*c1*m*Pi^2*n*d^5/(a*b)-
(26/81)*Q66b*m*Pi^2*n*d^3/(a*b)-
c2*((2186/15309)*Q12b*c1*m*Pi^2*. . .
    n*d^7/(a*b)-(242/1215)*Q12t*m*Pi^2*n*d^5/(a*b)-
(242/1215)*Q12b*m*Pi^2*n*d^5/(a*b)+(2186/15309)*Q12t*c1*m*Pi^2
*n*. . .
    d^7/(a*b)-(2/1215)*Q12m*m*Pi^2*n*d^5/(a*b)+(2/15309)*Q12m*
c1*m*Pi^2*n*d^7/(a*b));
K55=c2*(-(242/405)*Q44b*c1*d^5+(26/81)*Q44b*d^3-(2/405)*Q44m*c
1*d^5+(2/81)*Q44m*d^3-(242/405)*Q44t*c1*d^5+. . .
    (26/81)*Q44t*d^3)-c1*((2186/15309)*Q66b*c1*m^2*Pi^2*d^7/
a^2-(242/1215)*Q66t*m^2*Pi^2*d^5/a^2-(242/1215)*Q66b*. . .
    m^2*Pi^2*d^5/a^2+(2186/15309)*Q66t*c1*m^2*Pi^2*d^7/a^2-
(2/1215)*Q66m*m^2*Pi^2*d^5/a^2+(2/15309)*Q66m*c1*m^2*. . .
    Pi^2*d^7/a^2)+(242/1215)*Q22b*c1*n^2*Pi^2*d^5/b^2+(2/1215)
*Q22m*c1*n^2*Pi^2*d^5/b^2-(2/3)*Q44t*d+(242/1215)*. . .
    Q66b*c1*m^2*Pi^2*d^5/a^2-(26/81)*Q22b*n^2*Pi^2*d^3/b^2-
(2/81)*Q22m*n^2*Pi^2*d^3/b^2+(2/27)*Q44m*c1*d^3+. . .
    (242/1215)*Q22t*c1*n^2*Pi^2*d^5/b^2-
(26/81)*Q66t*m^2*Pi^2*d^3/a^2+(26/27)*Q44t*c1*d^3-
(26/81)*Q22t*n^2*Pi^2*. . .
    d^3/b^2-(2/3)*Q44b*d+(242/1215)*Q66t*c1*m^2*Pi^2*d^5/
a^2-(2/81)*Q66m*m^2*Pi^2*d^3/a^2-(2/3)*Q44m*d+(2/1215)*. . .
    Q66m*c1*m^2*Pi^2*d^5/a^2-c2*((2186/15309)*Q22b*c1*n^2*Pi^2
*d^7/b^2-(242/1215)*Q22t*n^2*Pi^2*d^5/b^2-(242/1215)*. . .
    Q22b*n^2*Pi^2*d^5/b^2+(2186/15309)*Q22t*c1*n^2*Pi^2*d^7/
b^2-(2/1215)*Q22m*n^2*Pi^2*d^5/b^2+(2/15309)*Q22m*. . .
    c1*n^2*Pi^2*d^7/b^2)+(26/27)*Q44b*c1*d^3-
(26/81)*Q66b*m^2*Pi^2*d^3/a^2;

KK=[K11 K12 K13 K14 K15;
    K21 K22 K23 K24 K25;
    K31 K32 K33 K34 K35;
    K41 K42 K43 K44 K45;
    K51 K52 K53 K54 K55];

KKG=[0 0 0 0 0;
    0 0 0 0 0;
    0 0 -alpha*n^2*Pi^2/b^2-m^2*Pi^2/a^2 0 0;
    0 0 0 0 0;
    0 0 0 0 0];
EQ1=KK+P*KKG;
EQ2=solve(det(EQ1));
BL(i)=(eval(abs(EQ2)))/(E1*d);
end
```

Appendix B

```
clc
clear
format long
%
N=10;
%----------------------------
en=.35;
uf=0;
ei=1;
Q=200;
%------------------------------------------------------------------
------------
for vi=1:Q
 vi
r1=11.43e-9;
r2=12.31e-9;
h=0.075e-9;
eta1=20;
E=1.8e12;
Length=eta1*r2;
rhof=1000;
rhot=2.25e3;
ks=0;
G=0;
Cm=0;
rf=0.5e-9;
d1=2*r1;
d2=2*r2;
df=2*rf;
A1=pi*((r1+h/2)^2-(r1-h/2)^2);
A2=pi*((r2+h/2)^2-(r2-h/2)^2);
Af=pi*r1^2;
I1=pi/4*((r1+h/2)^4-(r1-h/2)^4);
I2=pi/4*((r2+h/2)^4-(r2-h/2)^4);
If=pi/4*(r1-h/2)^4;
m1=rhot*A1;
m2=rhot*A2;
mf=rhof*Af;
T=0;
alpha=1.2e-6;
h11=0.95;
gama=50.0/sqrt(1e-15*E)*h11;
mub=3e-3;
c=9.513*2*r1*10e19;
cb1=c*Length^2/E/A1;
```

```
cb2=(r2/r1)^2*cb1;
f1=1;
f2=A2/A1;
rhob=rhof/rhot;
v=0.34;
lanx=E/(1-2*v)*alpha;
dTb=lanx*T/E;
Gpb=-0.00;
Kwb=-0.00;
G=E/2/(1+v);
lb1=(3*h)^2*G/4/E/Length^2;
lb2=(3*h)^2*G*A2/4/E/A1/Length^2;
Ib1=I1/A1/Length^2;
Ib2=I2/A1/Length^2;
Ibf=If/Af/rf^2;
G=E/2/(1+v);
xi1=(2*r1-h)/(2*r1+h);
xi2=(2*r2-h)/(2*r2+h);
Ks1=6*(1+v)*(1+xi1)^2/((7+6*v)*(1+xi1)^2+(20+12*v)*xi1^2);
Ks2=6*(1+v)*(1+xi2)^2/((7+6*v)*(1+xi2)^2+(20+12*v)*xi2^2);
beta1=Ks1*G/E;%beta1
beta2=Ks2*G*A2/E/A1;%beta2
%-------------------------------------------------------------
------------

for i=1:N
    xxx(i,1)=0.5*(1-cos(pi*(i-1)/(N-1)));
end
x=xxx;
x(1,1)=0;
x(2,1)=0.0001;
x(N-1,1)=0.9999;
x(N,1)=1;
%first derivative
C_1=zeros(N,N);
for i=1:N
    for j=1:N
        if j==i
            for k=1:N
                if k~=i
                    C_1(i,j)=C_1(i,j)+1/(x(i,1)-x(k,1));
                end
            end
        else
            NM=1;
            DN=1;
            for k=1:N
                if k~=j && k~=i
                    NM=NM*(x(i,1)-x(k,1));
                end
                if k~=j
```

```
                          DN=DN*(x(j,1)-x(k,1));
                 end
            end
            C_1(i,j)=NM/DN;
        end
    end
end
C_2=C_1*C_1;
C_3=C_1*C_2;
C_4=C_1*C_3;
C_5=C_1*C_4;
C_6=C_1*C_5;

%-------------------------
Kl1u1=1/(1-v^2)*C_2;
Kl1u2=zeros(N,N);
Kl1w1=zeros(N,N);
Kl1w2=zeros(N,N);
Kl1f1=zeros(N,N);
Kl1f2=zeros(N,N);
Kl1s1=3*gama*C_2+(en/gama)^2*gama*C_4;
Kl1s2=zeros(N,N);
%
Kl3u1=zeros(N,N);
Kl3u2=zeros(N,N);
Kl3w1=-dTb/(1-v^2)*C_2-(en/gama)^2*dTb/(1-v^2)*C_4+...
    beta1*C_2-lb1*C_4-cb1*eye(N)+mub*uf*C_1+(en/
gama)^2*f1*mub*uf/eta1*C_5...
    +dTb*C_2-uf^2*m1/mf*C_2+(en/gama)^2*f1*uf^2*C_4;
Kl3w2=cb1*eye(N)-(en/gama)^2*cb1*C_2;
Kl3f1=beta1*C_1+lb1*C_3+Gpb*C_1;
Kl3f2=zeros(N,N);
Kl3s1=zeros(N,N);
Kl3s2=zeros(N,N);
%
Kl5u1=zeros(N,N);
Kl5u2=zeros(N,N);
Kl5w1=-beta1*C_1-lb1*C_3;
Kl5w2=zeros(N,N);
Kl5f1=Ib1*C_2-beta1*eye(N)+lb1*C_2;
Kl5f2=zeros(N,N);
Kl5s1=zeros(N,N);
Kl5s2=zeros(N,N);
%
Kl7u1=gama*C_2-(en/gama)^2*gama*C_4;
Kl7u2=zeros(N,N);
Kl7w1=zeros(N,N);
Kl7w2=zeros(N,N);
Kl7f1=zeros(N,N);
Kl7f2=zeros(N,N);
Kl7s1=-C_2+2*(en/gama)^2*eta1*gama*C_4;
```

```
Kl7s2=zeros(N,N);
%
Kl2u1=zeros(N,N);
Kl2u2=1/(1-v^2)*C_2;
Kl2w1=zeros(N,N);
Kl2w2=zeros(N,N);
Kl2f1=zeros(N,N);
Kl2f2=zeros(N,N);
Kl2s1=zeros(N,N);
Kl2s2=3*gama*C_2+(en/gama)^2*gama*C_4;
%
Kl4u1=zeros(N,N);
Kl4u2=zeros(N,N);
Kl4w1=cb1*eye(N)-(en/gama)^2*cb2*C_2;
Kl4w2=beta2*C_2-dTb/(1-v^2)*C_2-(en/gama)^2*dTb/(1-v^2). . .
    *C_4-Kwb-(en/gama)^2*Kwb*C_2+Gpb*C_2+(en/
gama)^2*Gpb*C_4+. . .
    -lb2*C_4-cb1*eye(N)-(en/gama)^2*cb2*C_2;
Kl4f1=zeros(N,N);
Kl4f2=beta2*C_1+lb2*C_3;
Kl4s1=zeros(N,N);
Kl4s2=zeros(N,N);
%
Kl6u1=zeros(N,N);
Kl6u2=zeros(N,N);
Kl6w1=zeros(N,N);
Kl6w2=-beta2*C_1-lb2*C_3;
Kl6f1=zeros(N,N);
Kl6f2=Ib2/(1-v^2)*C_2-beta2*eye(N)+lb2*C_2;
Kl6s1=zeros(N,N);
Kl6s2=zeros(N,N);
%
Kl8u1=zeros(N,N);
Kl8u2=gama*C_2-(en/gama)^2*gama*C_4;
Kl8w1=zeros(N,N);
Kl8w2=zeros(N,N);
Kl8f1=zeros(N,N);
Kl8f2=zeros(N,N);
Kl8s1=zeros(N,N);
Kl8s2=-C_2+2*(en/gama)^2*eta1*C_4;
%
%
%
Ml1u1=-(30+rhob*f1)*(eye(N)-(en/gama)^2*C_2);
Ml1u2=zeros(N,N);
Ml1w1=zeros(N,N);
Ml1w2=zeros(N,N);
Ml1f1=zeros(N,N);
Ml1f2=zeros(N,N);
Ml1s1=zeros(N,N);
Ml1s2=zeros(N,N);
```

```
%
Ml3u1=zeros(N,N);
Ml3u2=zeros(N,N);
Ml3w1=-(30+f1*rhob)*eye(N)+(en/gama)^2*(1+f1*rhob)*C_2;
Ml3w2=zeros(N,N);
Ml3f1=zeros(N,N);
Ml3f2=zeros(N,N);
Ml3s1=zeros(N,N);
Ml3s2=zeros(N,N);
%
Ml5u1=zeros(N,N);
Ml5u2=zeros(N,N);
Ml5w1=zeros(N,N);
Ml5w2=zeros(N,N);
Ml5f1=-(Ib1+rhob*f1*Ibf)/eta1^3*eye(N)+(en/
gama)^2*(Ib1+rhob*f1*Ibf)/eta1^2*C_2;
Ml5f2=zeros(N,N);
Ml5s1=zeros(N,N);
Ml5s2=zeros(N,N);
%
Ml7u1=zeros(N,N);
Ml7u2=zeros(N,N);
Ml7w1=zeros(N,N);
Ml7w2=zeros(N,N);
Ml7f1=zeros(N,N);
Ml7f2=zeros(N,N);
Ml7s1=zeros(N,N);
Ml7s2=zeros(N,N);
%
Ml2u1=zeros(N,N);
Ml2u2=-eye(N)+(en/gama)^2*C_2;
Ml2w1=zeros(N,N);
Ml2w2=zeros(N,N);
Ml2f1=zeros(N,N);
Ml2f2=zeros(N,N);
Ml2s1=zeros(N,N);
Ml2s2=zeros(N,N);
%
Ml4u1=zeros(N,N);
Ml4u2=zeros(N,N);
Ml4w1=zeros(N,N);
Ml4w2=-eye(N)+(en/gama)^2*C_2;
Ml4f1=zeros(N,N);
Ml4f2=zeros(N,N);
Ml4s1=zeros(N,N);
Ml4s2=zeros(N,N);
%
Ml6u1=zeros(N,N);
Ml6u2=zeros(N,N);
Ml6w1=zeros(N,N);
Ml6w2=zeros(N,N);
```

```
Ml6f1=zeros(N,N);
Ml6f2=-Ib2/eta1^2*eye(N)+(en/gama)^2*Ib2/eta1^2*C_2;
Ml6s1=zeros(N,N);
Ml6s2=zeros(N,N);
%
Ml8u1=zeros(N,N);
Ml8u2=zeros(N,N);
Ml8w1=zeros(N,N);
Ml8w2=zeros(N,N);
Ml8f1=zeros(N,N);
Ml8f2=zeros(N,N);
Ml8s1=zeros(N,N);
Ml8s2=zeros(N,N);
%
Dl1u1=mub*sqrt(rhob)*f1/eta1*C_2/gama-(en/
gama)^2*sqrt(rhob)*mub*f1/eta1*C_4;
Dl1u2=zeros(N,N);
Dl1w1=zeros(N,N);
Dl1w2=zeros(N,N);
Dl1f1=zeros(N,N);
Dl1f2=zeros(N,N);
Dl1s1=zeros(N,N);
Dl1s2=zeros(N,N);
%
Dl3u1=zeros(N,N);
Dl3u2=zeros(N,N);
Dl3w1-2*f1*sqrt(rhob)*uf*C_1-2*(en/
gama)^2*f1*sqrt(rhob)*uf*C_3-. . .
     f1*sqrt(rhob)*mub/eta1*C_2+(en/gama)^2*f1*sqrt(rhob)*mub/
eta1*C_4;
Dl3w2=zeros(N,N);
Dl3f1=zeros(N,N);
Dl3f2=zeros(N,N);
Dl3s1=zeros(N,N);
Dl3s2=zeros(N,N);
%
Dl5u1=zeros(N,N);
Dl5u2=zeros(N,N);
Dl5w1=zeros(N,N);
Dl5w2=zeros(N,N);
Dl5f1=zeros(N,N);
Dl5f2=zeros(N,N);
Dl5s1=zeros(N,N);
Dl5s2=zeros(N,N);
%
Dl7u1=zeros(N,N);
Dl7w1=zeros(N,N);
Dl7u2=zeros(N,N);
Dl7w2=zeros(N,N);
Dl7f1=zeros(N,N);
Dl7f2=zeros(N,N);
```

```
Dl7s1=zeros(N,N);
Dl7s2=zeros(N,N);
%
Dl2u1=zeros(N,N);
Dl2u2=zeros(N,N);
Dl2w1=zeros(N,N);
Dl2w2=zeros(N,N);
Dl2f1=zeros(N,N);
Dl2f2=zeros(N,N);
Dl2s1=zeros(N,N);
Dl2s2=zeros(N,N);
%
Dl4u1=zeros(N,N);
Dl4u2=zeros(N,N);
Dl4w1=zeros(N,N);
Dl4w2=zeros(N,N);
Dl4f1=zeros(N,N);
Dl4f2=zeros(N,N);
Dl4s1=zeros(N,N);
Dl4s2=zeros(N,N);
%
Dl6u1=zeros(N,N);
Dl6u2=zeros(N,N);
Dl6w1=zeros(N,N);
Dl6w2=zeros(N,N);
Dl6f1=zeros(N,N);
Dl6f2=zeros(N,N);
Dl6s1=zeros(N,N);
Dl6s2=zeros(N,N);
%
Dl8u1=zeros(N,N);
Dl8u2=zeros(N,N);
Dl8w1=zeros(N,N);
Dl8w2=zeros(N,N);
Dl8f1=zeros(N,N);
Dl8f2=zeros(N,N);
Dl8s1=zeros(N,N);
Dl8s2=zeros(N,N);

%Boundary Conditions:

C_1_2=C_1(2,:);
C_1_N1=C_1(N-1,:);

Klw_2b=[C_1_2(1) C_1_2(2) C_1_2(N-1) C_1_2(N)];
Klw_2d=C_1_2(3:N-2);
Klw_N1b=[C_1_N1(1) C_1_N1(2) C_1_N1(N-1) C_1_N1(N)];
Klw_N1d=C_1_N1(3:N-2);

%-----------------------------------
bmKl1u1=[Kl1u1(2:N-1,1),Kl1u1(2:N-1,N)];
```

```
bmKl1u2=[Kl1u2(2:N-1,1),Kl1u2(2:N-1,N)];
bmKl1w1=[Kl1w1(2:N-1,1:2),Kl1w1(2:N-1,N-1:N)];
bmKl1w2=[Kl1w2(2:N-1,1:2),Kl1w2(2:N-1,N-1:N)];
bmKl1f1=[Kl1f1(2:N-1,1),Kl1f1(2:N-1,N)];
bmKl1f2=[Kl1f2(2:N-1,1),Kl1f2(2:N-1,N)];
bmKl1s1=[Kl1s1(2:N-1,1),Kl1s1(2:N-1,N)];
bmKl1s2=[Kl1s2(2:N-1,1),Kl1s2(2:N-1,N)];

bmKl2u1=[Kl2u1(2:N-1,1),Kl2u1(2:N-1,N)];
bmKl2u2=[Kl2u2(2:N-1,1),Kl2u2(2:N-1,N)];
bmKl2w1=[Kl2w1(2:N-1,1:2),Kl2w1(2:N-1,N-1:N)];
bmKl2w2=[Kl2w2(2:N-1,1:2),Kl2w2(2:N-1,N-1:N)];
bmKl2f1=[Kl2f1(2:N-1,1),Kl2f1(2:N-1,N)];
bmKl2f2=[Kl2f2(2:N-1,1),Kl2f2(2:N-1,N)];
bmKl2s1=[Kl2s1(2:N-1,1),Kl2s1(2:N-1,N)];
bmKl2s2=[Kl2s2(2:N-1,1),Kl2s2(2:N-1,N)];

bmKl3u1=[Kl3u1(3:N-2,1),Kl3u1(3:N-2,N)];
bmKl3u2=[Kl3u2(3:N-2,1),Kl3u2(3:N-2,N)];
bmKl3w1=[Kl3w1(3:N-2,1:2),Kl3w1(3:N-2,N-1:N)];
bmKl3w2=[Kl3w2(3:N-2,1:2),Kl3w2(3:N-2,N-1:N)];
bmKl3f1=[Kl3f1(3:N-2,1),Kl3f1(3:N-2,N)];
bmKl3f2=[Kl3f2(3:N-2,1),Kl3f2(3:N-2,N)];
bmKl3s1=[Kl3s1(3:N-2,1),Kl3s1(3:N-2,N)];
bmKl3s2=[Kl3s2(3:N-2,1),Kl3s2(3:N-2,N)];

bmKl4u1=[Kl4u1(3:N-2,1),Kl4u1(3:N-2,N)];
bmKl4u2=[Kl4u2(3:N-2,1),Kl4u2(3:N-2,N)];
bmKl4w1=[Kl4w1(3:N-2,1:2),Kl4w1(3:N-2,N-1:N)];
bmKl4w2=[Kl4w2(3:N-2,1:2),Kl4w2(3:N-2,N-1:N)];
bmKl4f1=[Kl4f1(3:N-2,1),Kl4f1(3:N-2,N)];
bmKl4f2=[Kl4f2(3:N-2,1),Kl4f2(3:N-2,N)];
bmKl4s1=[Kl4s1(3:N-2,1),Kl4s1(3:N-2,N)];
bmKl4s2=[Kl4s2(3:N-2,1),Kl4s2(3:N-2,N)];

bmKl5u1=[Kl5u1(2:N-1,1),Kl5u1(2:N-1,N)];
bmKl5u2=[Kl5u2(2:N-1,1),Kl5u2(2:N-1,N)];
bmKl5w1=[Kl5w1(2:N-1,1:2),Kl5w1(2:N-1,N-1:N)];
bmKl5w2=[Kl5w2(2:N-1,1:2),Kl5w2(2:N-1,N-1:N)];
bmKl5f1=[Kl5f1(2:N-1,1),Kl5f1(2:N-1,N)];
bmKl5f2=[Kl5f2(2:N-1,1),Kl5f2(2:N-1,N)];
bmKl5s1=[Kl5s1(2:N-1,1),Kl5s1(2:N-1,N)];
bmKl5s2=[Kl5s2(2:N-1,1),Kl5s2(2:N-1,N)];

bmKl6u1=[Kl6u1(2:N-1,1),Kl6u1(2:N-1,N)];
bmKl6u2=[Kl6u2(2:N-1,1),Kl6u2(2:N-1,N)];
bmKl6w1=[Kl6w1(2:N-1,1:2),Kl6w1(2:N-1,N-1:N)];
bmKl6w2=[Kl6w2(2:N-1,1:2),Kl6w2(2:N-1,N-1:N)];
bmKl6f1=[Kl6f1(2:N-1,1),Kl6f1(2:N-1,N)];
bmKl6f2=[Kl6f2(2:N-1,1),Kl6f2(2:N-1,N)];
bmKl6s1=[Kl6s1(2:N-1,1),Kl6s1(2:N-1,N)];
```

```
bmKl6s2=[Kl6s2(2:N-1,1),Kl6s2(2:N-1,N)];

bmKl7u1=[Kl7u1(2:N-1,1),Kl7u1(2:N-1,N)];
bmKl7u2=[Kl7u2(2:N-1,1),Kl7u2(2:N-1,N)];
bmKl7w1=[Kl7w1(2:N-1,1:2),Kl7w1(2:N-1,N-1:N)];
bmKl7w2=[Kl7w2(2:N-1,1:2),Kl7w2(2:N-1,N-1:N)];
bmKl7f1=[Kl7f1(2:N-1,1),Kl7f1(2:N-1,N)];
bmKl7f2=[Kl7f2(2:N-1,1),Kl7f2(2:N-1,N)];
bmKl7s1=[Kl7s1(2:N-1,1),Kl7s1(2:N-1,N)];
bmKl7s2=[Kl7s2(2:N-1,1),Kl7s2(2:N-1,N)];

bmKl8u1=[Kl8u1(2:N-1,1),Kl8u1(2:N-1,N)];
bmKl8u2=[Kl8u2(2:N-1,1),Kl8u2(2:N-1,N)];
bmKl8w1=[Kl8w1(2:N-1,1:2),Kl8w1(2:N-1,N-1:N)];
bmKl8w2=[Kl8w2(2:N-1,1:2),Kl8w2(2:N-1,N-1:N)];
bmKl8f1=[Kl8f1(2:N-1,1),Kl8f1(2:N-1,N)];
bmKl8f2=[Kl8f2(2:N-1,1),Kl8f2(2:N-1,N)];
bmKl8s1=[Kl8s1(2:N-1,1),Kl8s1(2:N-1,N)];
bmKl8s2=[Kl8s2(2:N-1,1),Kl8s2(2:N-1,N)];

%-------------------------------------------
%-------------------------------------------

bmMl1u1=[Ml1u1(2:N-1,1),Ml1u1(2:N-1,N)];
bmMl1u2=[Ml1u2(2:N-1,1),Ml1u2(2:N-1,N)];
bmMl1w1=[Ml1w1(2:N-1,1:2),Ml1w1(2:N-1,N-1:N)];
bmMl1w2=[Ml1w2(2:N-1,1:2),Ml1w2(2:N-1,N-1:N)];
bmMl1f1=[Ml1f1(2:N-1,1),Ml1f1(2:N-1,N)];
bmMl1f2=[Ml1f2(2:N-1,1),Ml1f2(2:N-1,N)];
bmMl1s1=[Ml1s1(2:N-1,1),Ml1s1(2:N-1,N)];
bmMl1s2=[Ml1s2(2:N-1,1),Ml1s2(2:N-1,N)];

bmMl2u1=[Ml2u1(2:N-1,1),Ml2u1(2:N-1,N)];
bmMl2u2=[Ml2u2(2:N-1,1),Ml2u2(2:N-1,N)];
bmMl2w1=[Ml2w1(2:N-1,1:2),Ml2w1(2:N-1,N-1:N)];
bmMl2w2=[Ml2w2(2:N-1,1:2),Ml2w2(2:N-1,N-1:N)];
bmMl2f1=[Ml2f1(2:N-1,1),Ml2f1(2:N-1,N)];
bmMl2f2=[Ml2f2(2:N-1,1),Ml2f2(2:N-1,N)];
bmMl2s1=[Ml2s1(2:N-1,1),Ml2s1(2:N-1,N)];
bmMl2s2=[Ml2s2(2:N-1,1),Ml2s2(2:N-1,N)];

bmMl3u1=[Ml3u1(3:N-2,1),Ml3u1(3:N-2,N)];
bmMl3u2=[Ml3u2(3:N-2,1),Ml3u2(3:N-2,N)];
bmMl3w1=[Ml3w1(3:N-2,1:2),Ml3w1(3:N-2,N-1:N)];
bmMl3w2=[Ml3w2(3:N-2,1:2),Ml3w2(3:N-2,N-1:N)];
bmMl3f1=[Ml3f1(3:N-2,1),Ml3f1(3:N-2,N)];
bmMl3f2=[Ml3f2(3:N-2,1),Ml3f2(3:N-2,N)];
bmMl3s1=[Ml3s1(3:N-2,1),Ml3s1(3:N-2,N)];
bmMl3s2=[Ml3s2(3:N-2,1),Ml3s2(3:N-2,N)];

bmMl4u1=[Ml4u1(3:N-2,1),Ml4u1(3:N-2,N)];
```

```
bmM14u2=[M14u2(3:N-2,1),M14u2(3:N-2,N)];
bmM14w1=[M14w1(3:N-2,1:2),M14w1(3:N-2,N-1:N)];
bmM14w2=[M14w2(3:N-2,1:2),M14w2(3:N-2,N-1:N)];
bmM14f1=[M14f1(3:N-2,1),M14f1(3:N-2,N)];
bmM14f2=[M14f2(3:N-2,1),M14f2(3:N-2,N)];
bmM14s1=[M14s1(3:N-2,1),M14s1(3:N-2,N)];
bmM14s2=[M14s2(3:N-2,1),M14s2(3:N-2,N)];

bmM15u1=[M15u1(2:N-1,1),M15u1(2:N-1,N)];
bmM15u2=[M15u2(2:N-1,1),M15u2(2:N-1,N)];
bmM15w1=[M15w1(2:N-1,1:2),M15w1(2:N-1,N-1:N)];
bmM15w2=[M15w2(2:N-1,1:2),M15w2(2:N-1,N-1:N)];
bmM15f1=[M15f1(2:N-1,1),M15f1(2:N-1,N)];
bmM15f2=[M15f2(2:N-1,1),M15f2(2:N-1,N)];
bmM15s1=[M15s1(2:N-1,1),M15s1(2:N-1,N)];
bmM15s2=[M15s2(2:N-1,1),M15s2(2:N-1,N)];

bmM16u1=[M16u1(2:N-1,1),M16u1(2:N-1,N)];
bmM16u2=[M16u2(2:N-1,1),M16u2(2:N-1,N)];
bmM16w1=[M16w1(2:N-1,1:2),M16w1(2:N-1,N-1:N)];
bmM16w2=[M16w2(2:N-1,1:2),M16w2(2:N-1,N-1:N)];
bmM16f1=[M16f1(2:N-1,1),M16f1(2:N-1,N)];
bmM16f2=[M16f2(2:N-1,1),M16f2(2:N-1,N)];
bmM16s1=[M16s1(2:N-1,1),M16s1(2:N-1,N)];
bmM16s2=[M16s2(2:N-1,1),M16s2(2:N-1,N)];

bmM17u1=[M17u1(2:N-1,1),M17u1(2:N-1,N)];
bmM17u2=[M17u2(2:N-1,1),M17u2(2:N-1,N)];
bmM17w1=[M17w1(2:N-1,1:2),M17w1(2:N-1,N-1:N)];
bmM17w2=[M17w2(2:N-1,1:2),M17w2(2:N-1,N-1:N)];
bmM17f1=[M17f1(2:N-1,1),M17f1(2:N-1,N)];
bmM17f2=[M17f2(2:N-1,1),M17f2(2:N-1,N)];
bmM17s1=[M17s1(2:N-1,1),M17s1(2:N-1,N)];
bmM17s2=[M17s2(2:N-1,1),M17s2(2:N-1,N)];

bmM18u1=[M18u1(2:N-1,1),M18u1(2:N-1,N)];
bmM18u2=[M18u2(2:N-1,1),M18u2(2:N-1,N)];
bmM18w1=[M18w1(2:N-1,1:2),M18w1(2:N-1,N-1:N)];
bmM18w2=[M18w2(2:N-1,1:2),M18w2(2:N-1,N-1:N)];
bmM18f1=[M18f1(2:N-1,1),M18f1(2:N-1,N)];
bmM18f2=[M18f2(2:N-1,1),M18f2(2:N-1,N)];
bmM18s1=[M18s1(2:N-1,1),M18s1(2:N-1,N)];
bmM18s2=[M18s2(2:N-1,1),M18s2(2:N-1,N)];
%----------------------------------------
%----------------------------------------

bmD11u1=[D11u1(2:N-1,1),D11u1(2:N-1,N)];
bmD11u2=[D11u2(2:N-1,1),D11u2(2:N-1,N)];
bmD11w1=[D11w1(2:N-1,1:2),D11w1(2:N-1,N-1:N)];
bmD11w2=[D11w2(2:N-1,1:2),D11w2(2:N-1,N-1:N)];
bmD11f1=[D11f1(2:N-1,1),D11f1(2:N-1,N)];
```

```
bmD11f2=[D11f2(2:N-1,1),D11f2(2:N-1,N)];
bmD11s1=[D11s1(2:N-1,1),D11s1(2:N-1,N)];
bmD11s2=[D11s2(2:N-1,1),D11s2(2:N-1,N)];

bmD12u1=[D12u1(2:N-1,1),D12u1(2:N-1,N)];
bmD12u2=[D12u2(2:N-1,1),D12u2(2:N-1,N)];
bmD12w1=[D12w1(2:N-1,1:2),D12w1(2:N-1,N-1:N)];
bmD12w2=[D12w2(2:N-1,1:2),D12w2(2:N-1,N-1:N)];
bmD12f1=[D12f1(2:N-1,1),D12f1(2:N-1,N)];
bmD12f2=[D12f2(2:N-1,1),D12f2(2:N-1,N)];
bmD12s1=[D12s1(2:N-1,1),D12s1(2:N-1,N)];
bmD12s2=[D12s2(2:N-1,1),D12s2(2:N-1,N)];

bmD13u1=[D13u1(3:N-2,1),D13u1(3:N-2,N)];
bmD13u2=[D13u2(3:N-2,1),D13u2(3:N-2,N)];
bmD13w1=[D13w1(3:N-2,1:2),D13w1(3:N-2,N-1:N)];
bmD13w2=[D13w2(3:N-2,1:2),D13w2(3:N-2,N-1:N)];
bmD13f1=[D13f1(3:N-2,1),D13f1(3:N-2,N)];
bmD13f2=[D13f2(3:N-2,1),D13f2(3:N-2,N)];
bmD13s1=[D13s1(3:N-2,1),D13s1(3:N-2,N)];
bmD13s2=[D13s2(3:N-2,1),D13s2(3:N-2,N)];

bmD14u1=[D14u1(3:N-2,1),D14u1(3:N-2,N)];
bmD14u2=[D14u2(3:N-2,1),D14u2(3:N-2,N)];
bmD14w1=[D14w1(3:N-2,1:2),D14w1(3:N-2,N-1:N)];
bmD14w2=[D14w2(3:N-2,1:2),D14w2(3:N-2,N-1:N)];
bmD14f1=[D14f1(3:N-2,1),D14f1(3:N-2,N)];
bmD14f2=[D14f2(3:N-2,1),D14f2(3:N-2,N)];
bmD14s1=[D14s1(3:N-2,1),D14s1(3:N-2,N)];
bmD14s2=[D14s2(3:N-2,1),D14s2(3:N-2,N)];

bmD15u1=[D15u1(2:N-1,1),D15u1(2:N-1,N)];
bmD15u2=[D15u2(2:N-1,1),D15u2(2:N-1,N)];
bmD15w1=[D15w1(2:N-1,1:2),D15w1(2:N-1,N-1:N)];
bmD15w2=[D15w2(2:N-1,1:2),D15w2(2:N-1,N-1:N)];
bmD15f1=[D15f1(2:N-1,1),D15f1(2:N-1,N)];
bmD15f2=[D15f2(2:N-1,1),D15f2(2:N-1,N)];
bmD15s1=[D15s1(2:N-1,1),D15s1(2:N-1,N)];
bmD15s2=[D15s2(2:N-1,1),D15s2(2:N-1,N)];

bmD16u1=[D16u1(2:N-1,1),D16u1(2:N-1,N)];
bmD16u2=[D16u2(2:N-1,1),D16u2(2:N-1,N)];
bmD16w1=[D16w1(2:N-1,1:2),D16w1(2:N-1,N-1:N)];
bmD16w2=[D16w2(2:N-1,1:2),D16w2(2:N-1,N-1:N)];
bmD16f1=[D16f1(2:N-1,1),D16f1(2:N-1,N)];
bmD16f2=[D16f2(2:N-1,1),D16f2(2:N-1,N)];
bmD16s1=[D16s1(2:N-1,1),D16s1(2:N-1,N)];
bmD16s2=[D16s2(2:N-1,1),D16s2(2:N-1,N)];

bmD17u1=[D17u1(2:N-1,1),D17u1(2:N-1,N)];
bmD17u2=[D17u2(2:N-1,1),D17u2(2:N-1,N)];
```

```
bmD17w1=[D17w1(2:N-1,1:2),D17w1(2:N-1,N-1:N)];
bmD17w2=[D17w2(2:N-1,1:2),D17w2(2:N-1,N-1:N)];
bmD17f1=[D17f1(2:N-1,1),D17f1(2:N-1,N)];
bmD17f2=[D17f2(2:N-1,1),D17f2(2:N-1,N)];
bmD17s1=[D17s1(2:N-1,1),D17s1(2:N-1,N)];
bmD17s2=[D17s2(2:N-1,1),D17s2(2:N-1,N)];


bmD18u1=[D18u1(2:N-1,1),D18u1(2:N-1,N)];
bmD18u2=[D18u2(2:N-1,1),D18u2(2:N-1,N)];
bmD18w1=[D18w1(2:N-1,1:2),D18w1(2:N-1,N-1:N)];
bmD18w2=[D18w2(2:N-1,1:2),D18w2(2:N-1,N-1:N)];
bmD18f1=[D18f1(2:N-1,1),D18f1(2:N-1,N)];
bmD18f2=[D18f2(2:N-1,1),D18f2(2:N-1,N)];
bmD18s1=[D18s1(2:N-1,1),D18s1(2:N-1,N)];
bmD18s2=[D18s2(2:N-1,1),D18s2(2:N-1,N)];
%-------------------------------------
%-------------------------------------
mK11u1=K11u1(2:N-1,2:N-1);
mK11u2=K11u2(2:N-1,2:N-1);
mK11w1=K11w1(2:N-1,3:N-2);
mK11w2=K11w2(2:N-1,3:N-2);
mK11f1=K11f1(2:N-1,2:N-1);
mK11f2=K11f2(2:N-1,2:N-1);
mK11s1=K11s1(2:N-1,2:N-1);
mK11s2=K11s2(2:N-1,2:N-1);

mK12u1=K12u1(2:N-1,2:N-1);
mK12u2=K12u2(2:N-1,2:N-1);
mK12w1=K12w1(2:N-1,3:N-2);
mK12w2=K12w2(2:N-1,3:N-2);
mK12f1=K12f1(2:N-1,2:N-1);
mK12f2=K12f2(2:N-1,2:N-1);
mK12s1=K12s1(2:N-1,2:N-1);
mK12s2=K12s2(2:N-1,2:N-1);

mK13u1=K13u1(3:N-2,2:N-1);
mK13u2=K13u2(3:N-2,2:N-1);
mK13w1=K13w1(3:N-2,3:N-2);
mK13w2=K13w2(3:N-2,3:N-2);
mK13f1=K13f1(3:N-2,2:N-1);
mK13f2=K13f2(3:N-2,2:N-1);
mK13s1=K13s1(3:N-2,2:N-1);
mK13s2=K13s2(3:N-2,2:N-1);

mK14u1=K14u1(3:N-2,2:N-1);
mK14u2=K14u2(3:N-2,2:N-1);
mK14w1=K14w1(3:N-2,3:N-2);
mK14w2=K14w2(3:N-2,3:N-2);
mK14f1=K14f1(3:N-2,2:N-1);
mK14f2=K14f2(3:N-2,2:N-1);
mK14s1=K14s1(3:N-2,2:N-1);
```

```
mK14s2=K14s2(3:N-2,2:N-1);

mK15u1=K15u1(2:N-1,?:N-1);
mK15u2=K15u2(2:N-1,2:N-1);
mK15w1=K15w1(2:N-1,3:N-2);
mK15w2=K15w2(2:N-1,3:N-2);
mK15f1=K15f1(2:N-1,2:N-1);
mK15f2=K15f2(2:N-1,2:N-1);
mK15s1=K15s1(2:N-1,2:N-1);
mK15s2=K15s2(2:N-1,2:N-1);

mK16u1=K16u1(2:N-1,2:N-1);
mK16u2=K16u2(2:N-1,2:N-1);
mK16w1=K16w1(2:N-1,3:N-2);
mK16w2=K16w2(2:N-1,3:N-2);
mK16f1=K16f1(2:N-1,2:N-1);
mK16f2=K16f2(2:N-1,2:N-1);
mK16s1=K16s1(2:N-1,2:N-1);
mK16s2=K16s2(2:N-1,2:N-1);

mK17u1=K17u1(2:N-1,2:N-1);
mK17u2=K17u2(2:N-1,2:N-1);
mK17w1=K17w1(2:N-1,3:N-2);
mK17w2=K17w2(2:N-1,3:N-2);
mK17f1=K17f1(2:N-1,2:N-1);
mK17f2=K17f2(2:N-1,2:N-1);
mK17s1=K17s1(2:N-1,2:N-1);
mK17s2=K17s2(2:N-1,2:N-1);

mK18u1=K18u1(2:N-1,2:N-1);
mK18u2=K18u2(2:N-1,2:N-1);
mK18w1=K18w1(2:N-1,3:N-2);
mK18w2=K18w2(2:N-1,3:N-2);
mK18f1=K18f1(2:N-1,2:N-1);
mK18f2=K18f2(2:N-1,2:N-1);
mK18s1=K18s1(2:N-1,2:N-1);
mK18s2=K18s2(2:N-1,2:N-1);
%--------------------------------
%--------------------------------

mM11u1=M11u1(2:N-1,2:N-1);
mM11u2=M11u2(2:N-1,2:N-1);
mM11w1=M11w1(2:N-1,3:N-2);
mM11w2=M11w2(2:N-1,3:N-2);
mM11f1=M11f1(2:N-1,2:N-1);
mM11f2=M11f2(2:N-1,2:N-1);
mM11s1=M11s1(2:N-1,2:N-1);
mM11s2=M11s2(2:N-1,2:N-1);

mM12u1=M12u1(2:N-1,2:N-1);
mM12u2=M12u2(2:N-1,2:N-1);
```

```
mMl2w1=Ml2w1(2:N-1,3:N-2);
mMl2w2=Ml2w2(2:N-1,3:N-2);
mMl2f1=Ml2f1(2:N-1,2:N-1);
mMl2f2=Ml2f2(2:N-1,2:N-1);
mMl2s1=Ml2s1(2:N-1,2:N-1);
mMl2s2=Ml2s2(2:N-1,2:N-1);


mMl3u1=Ml3u1(3:N-2,2:N-1);
mMl3u2=Ml3u2(3:N-2,2:N-1);
mMl3w1=Ml3w1(3:N-2,3:N-2);
mMl3w2=Ml3w2(3:N-2,3:N-2);
mMl3f1=Ml3f1(3:N-2,2:N-1);
mMl3f2=Ml3f2(3:N-2,2:N-1);
mMl3s1=Ml3s1(3:N-2,2:N-1);
mMl3s2=Ml3s2(3:N-2,2:N-1);


mMl4u1=Ml4u1(3:N-2,2:N-1);
mMl4u2=Ml4u2(3:N-2,2:N-1);
mMl4w1=Ml4w1(3:N-2,3:N-2);
mMl4w2=Ml4w2(3:N-2,3:N-2);
mMl4f1=Ml4f1(3:N-2,2:N-1);
mMl4f2=Ml4f2(3:N-2,2:N-1);
mMl4s1=Ml4s1(3:N-2,2:N-1);
mMl4s2=Ml4s2(3:N-2,2:N-1);


mMl5u1=Ml5u1(2:N-1,2:N-1);
mMl5u2=Ml5u2(2:N-1,2:N-1);
mMl5w1=Ml5w1(2:N-1,3:N-2);
mMl5w2=Ml5w2(2:N-1,3:N-2);
mMl5f1=Ml5f1(2:N-1,2:N-1);
mMl5f2=Ml5f2(2:N-1,2:N-1);
mMl5s1=Ml5s1(2:N-1,2:N-1);
mMl5s2=Ml5s2(2:N-1,2:N-1);


mMl6u1=Ml6u1(2:N-1,2:N-1);
mMl6u2=Ml6u2(2:N-1,2:N-1);
mMl6w1=Ml6w1(2:N-1,3:N-2);
mMl6w2=Ml6w2(2:N-1,3:N-2);
mMl6f1=Ml6f1(2:N-1,2:N-1);
mMl6f2=Ml6f2(2:N-1,2:N-1);
mMl6s1=Ml6s1(2:N-1,2:N-1);
mMl6s2=Ml6s2(2:N-1,2:N-1);


mMl7u1=Ml7u1(2:N-1,2:N-1);
mMl7u2=Ml7u2(2:N-1,2:N-1);
mMl7w1=Ml7w1(2:N-1,3:N-2);
mMl7w2=Ml7w2(2:N-1,3:N-2);
mMl7f1=Ml7f1(2:N-1,2:N-1);
mMl7f2=Ml7f2(2:N-1,2:N-1);
mMl7s1=Ml7s1(2:N-1,2:N-1);
mMl7s2=Ml7s2(2:N-1,2:N-1);
```

```
mM18u1=M18u1(2:N-1,2:N-1);
mM18u2=M18u2(2:N-1,2:N-1);
mM18w1=M18w1(2:N-1,3:N-2);
mM18w2=M18w2(2:N-1,3:N-2);
mM18f1=M18f1(2:N-1,2:N-1);
mM18f2=M18f2(2:N-1,2:N-1);
mM18s1=M18s1(2:N-1,2:N-1);
mM18s2=M18s2(2:N-1,2:N-1);
%----------------------------
%----------------------------

mDl1u1=Dl1u1(2:N-1,2:N-1);
mDl1u2=Dl1u2(2:N-1,2:N-1);
mDl1w1=Dl1w1(2:N-1,3:N-2);
mDl1w2=Dl1w2(2:N-1,3:N-2);
mDl1f1=Dl1f1(2:N-1,2:N-1);
mDl1f2=Dl1f2(2:N-1,2:N-1);
mDl1s1=Dl1s1(2:N-1,2:N-1);
mDl1s2=Dl1s2(2:N-1,2:N-1);

mDl2u1=Dl2u1(2:N-1,2:N-1);
mDl2u2=Dl2u2(2:N-1,2:N-1);
mDl2w1=Dl2w1(2:N-1,3:N-2);
mDl2w2=Dl2w2(2:N-1,3:N-2);
mDl2f1=Dl2f1(2:N-1,2:N-1);
mDl2f2=Dl2f2(2:N-1,2:N-1);
mDl2s1=Dl2s1(2:N-1,2:N-1);
mDl2s2=Dl2s2(2:N-1,2:N-1);

mDl3u1=Dl3u1(3:N-2,2:N-1);
mDl3u2=Dl3u2(3:N-2,2:N-1);
mDl3w1=Dl3w1(3:N-2,3:N-2);
mDl3w2=Dl3w2(3:N-2,3:N-2);
mDl3f1=Dl3f1(3:N-2,2:N-1);
mDl3f2=Dl3f2(3:N-2,2:N-1);
mDl3s1=Dl3s1(3:N-2,2:N-1);
mDl3s2=Dl3s2(3:N-2,2:N-1);

mDl4u1=Dl4u1(3:N-2,2:N-1);
mDl4u2=Dl4u2(3:N-2,2:N-1);
mDl4w1=Dl4w1(3:N-2,3:N-2);
mDl4w2=Dl4w2(3:N-2,3:N-2);
mDl4f1=Dl4f1(3:N-2,2:N-1);
mDl4f2=Dl4f2(3:N-2,2:N-1);
mDl4s1=Dl4s1(3:N-2,2:N-1);
mDl4s2=Dl4s2(3:N-2,2:N-1);

mDl5u1=Dl5u1(2:N-1,2:N-1);
mDl5u2=Dl5u2(2:N-1,2:N-1);
mDl5w1=Dl5w1(2:N-1,3:N-2);
mDl5w2=Dl5w2(2:N-1,3:N-2);
```

```
mD15f1=D15f1(2:N-1,2:N-1);
mD15f2=D15f2(2:N-1,2:N-1);
mD15s1=D15s1(2:N-1,2:N-1);
mD15s2=D15s2(2:N-1,2:N-1);

mD16u1=D16u1(2:N-1,2:N-1);
mD16u2=D16u2(2:N-1,2:N-1);
mD16w1=D16w1(2:N-1,3:N-2);
mD16w2=D16w2(2:N-1,3:N-2);
mD16f1=D16f1(2:N-1,2:N-1);
mD16f2=D16f2(2:N-1,2:N-1);
mD16s1=D16s1(2:N-1,2:N-1);
mD16s2=D16s2(2:N-1,2:N-1);

mD17u1=D17u1(2:N-1,2:N-1);
mD17u2=D17u2(2:N-1,2:N-1);
mD17w1=D17w1(2:N-1,3:N-2);
mD17w2=D17w2(2:N-1,3:N-2);
mD17f1=D17f1(2:N-1,2:N-1);
mD17f2=D17f2(2:N-1,2:N-1);
mD17s1=D17s1(2:N-1,2:N-1);
mD17s2=D17s2(2:N-1,2:N-1);

mD18u1=D18u1(2:N-1,2:N-1);
mD18u2=D18u2(2:N-1,2:N-1);
mD18w1=D18w1(2:N-1,3:N-2);
mD18w2=D18w2(2:N-1,3:N-2);
mD18f1=D18f1(2:N-1,2:N-1);
mD18f2=D18f2(2:N-1,2:N-1);
mD18s1=D18s1(2:N-1,2:N-1);
mD18s2=D18s2(2:N-1,2:N-1);

Kl=[1 zeros(1,8*N-1);. . .
    0 1 zeros(1,8*N-2);. . .
    %-----------------------------------
    zeros(1,2) 1 zeros(1,8*N-2-1);. . .
    zeros(1,2+1) 1 zeros(1,8*N-2*2);. . .
    %-----------------------------------
    zeros(1,2*2) 1 zeros(1,8*N-2*2-1);. . .
    zeros(1,2*2),Klw_2b,zeros(1,1*4),zeros(1,4*2),zeros(1,2*
(N-2)),Klw_2d,zeros(1,1*(N-4)),zeros(1,4*(N-2));. . .
    zeros(1,2*2),Klw_N1b,zeros(1,1*4),zeros(1,4*2),zeros(1,2*
(N-2)),Klw_N1d,zeros(1,1*(N-4)),zeros(1,4*(N-2));. . .
    zeros(1,2*2+3) 1 zeros(1,8*N-2*2-4);. . .
    %-----------------------------------
    zeros(1,2*2+1*4) 1 zeros(1,8*N-2*2-1*4-1);. . .
    zeros(1,2*2+1*4),Klw_2b,zeros(1,4*2),zeros(1,2*(N-
2)),zeros(1,1*(N-4)),Klw_2d,zeros(1,4*(N-2));. . .
    zeros(1,2*2+1*4),Klw_N1b,zeros(1,4*2),zeros(1,2*(N-
2)),zeros(1,1*(N-4)),Klw_N1d,zeros(1,4*(N-2));. . .
    zeros(1,2*2+1*4+3) 1 zeros(1,8*N-2*2-2*4);. . .
```

```
%------------------------------------
zeros(1,2*2+2*4) 1 zeros(1,8*N-2*2-2*4-1);. . .
zeros(1,2*2+2*4+1) 1 zeros(1,8*N-3*2-2*4);. . .
%------------------------------------
zeros(1,3*2+2*4) 1 zeros(1,8*N-3*2-2*4-1);. . .
zeros(1,3*2+2*4+1) 1 zeros(1,8*N-4*2-2*4);. . .
%------------------------------------
zeros(1,4*2+2*4) 1 zeros(1,8*N-4*2-2*4-1);. . .
zeros(1,4*2+2*4+1) 1 zeros(1,8*N-5*2-2*4);. . .
%------------------------------------
zeros(1,5*2+2*4) 1 zeros(1,8*N-5*2-2*4-1);. . .
zeros(1,5*2+2*4+1) 1 zeros(1,8*N-6*2-2*4);. . .
%------------------------------------
    bmKl1u1,bmKl1u2,bmKl1w1,bmKl1w2,bmKl1f1,bmKl1f2,bmKl1s1,
bmKl1s2,mKl1u1,mKl1u2,mKl1w1,mKl1w2,mKl1f1,mKl1f2,mKl1s1,
mKl1s2;. . .
    bmKl2u1,bmKl2u2,bmKl2w1,bmKl2w2,bmKl2f1,bmKl2f2,bmKl2s1,
bmKl2s2,mKl2u1,mKl2u2,mKl2w1,mKl2w2,mKl2f1,mKl2f2,mKl2s1,
mKl2s2;. . .
    bmKl3u1,bmKl3u2,bmKl3w1,bmKl3w2,bmKl3f1,bmKl3f2,bmKl3s1,
bmKl3s2,mKl3u1,mKl3u2,mKl3w1,mKl3w2,mKl3f1,mKl3f2,mKl3s1,
mKl3s2;. . .
    bmKl4u1,bmKl4u2,bmKl4w1,bmKl4w2,bmKl4f1,bmKl4f2,bmKl4s1,
bmKl4s2,mKl4u1,mKl4u2,mKl4w1,mKl4w2,mKl4f1,mKl4f2,mKl4s1,
mKl4s2;. . .
    bmKl5u1,bmKl5u2,bmKl5w1,bmKl5w2,bmKl5f1,bmKl5f2,bmKl5s1,
bmKl5s2,mKl5u1,mKl5u2,mKl5w1,mKl5w2,mKl5f1,mKl5f2,mKl5s1,
mKl5s2;. . .
    bmKl6u1,bmKl6u2,bmKl6w1,bmKl6w2,bmKl6f1,bmKl6f2,bmKl6s1,
bmKl6s2,mKl6u1,mKl6u2,mKl6w1,mKl6w2,mKl6f1,mKl6f2,mKl6s1,
mKl6s2;. . .
    bmKl7u1,bmKl7u2,bmKl7w1,bmKl7w2,bmKl7f1,bmKl7f2,bmKl7s1,
bmKl7s2,mKl7u1,mKl7u2,mKl7w1,mKl7w2,mKl7f1,mKl7f2,mKl7s1,
mKl7s2;. . .
    bmKl8u1,bmKl8u2,bmKl8w1,bmKl8w2,bmKl8f1,bmKl8f2,bmKl8s1,
bmKl8s2,mKl8u1,mKl8u2,mKl8w1,mKl8w2,mKl8f1,mKl8f2,mKl8s1,
mKl8s2];

Ml=[
zeros(1,8*N)                                                  ;. . .

zeros(1,8*N)                                                  ;. . .

zeros(1,8*N)                                                  ;. . .

zeros(1,8*N)                                                  ;. . .

zeros(1,8*N)                                                  ;. . .

zeros(1,8*N)                                                  ;. . .
```

```
zeros(1,8*N)                                                    ; . . .

zeros(1,8*N)                                                    ; . . .

zeros(1,8*N)                                                    ; . . .

zeros(1,8*N)                                                    ; . . .

zeros(1,8*N)                                                    ; . . .

zeros(1,8*N)                                                    ; . . .

zeros(1,8*N)                                                    ; . . .

zeros(1,8*N)                                                    ; . . .

zeros(1,8*N)                                                    ; . . .

zeros(1,8*N)                                                    ; . . .

zeros(1,8*N)                                                    ; . . .

zeros(1,8*N)                                                    ; . . .

zeros(1,8*N)                                                    ; . . .

zeros(1,8*N)                                                    ; . . .
    bmMl1u1,bmMl1u2,bmMl1w1,bmMl1w2,bmMl1f1,bmMl1f2,bmMl1s1,
bmMl1s2,mMl1u1,mMl1u2,mMl1w1,mMl1w2,mMl1f1,mMl1f2,mMl1s1,
mMl1s2;. . .
    bmMl2u1,bmMl2u2,bmMl2w1,bmMl2w2,bmMl2f1,bmMl2f2,bmMl2s1,
bmMl2s2,mMl2u1,mMl2u2,mMl2w1,mMl2w2,mMl2f1,mMl2f2,mMl2s1,
mMl2s2;. . .
    bmMl3u1,bmMl3u2,bmMl3w1,bmMl3w2,bmMl3f1,bmMl3f2,bmMl3s1,
bmMl3s2,mMl3u1,mMl3u2,mMl3w1,mMl3w2,mMl3f1,mMl3f2,mMl3s1,
mMl3s2;. . .
    bmMl4u1,bmMl4u2,bmMl4w1,bmMl4w2,bmMl4f1,bmMl4f2,bmMl4s1,
bmMl4s2,mMl4u1,mMl4u2,mMl4w1,mMl4w2,mMl4f1,mMl4f2,mMl4s1,
mMl4s2;. . .
    bmMl5u1,bmMl5u2,bmMl5w1,bmMl5w2,bmMl5f1,bmMl5f2,bmMl5s1,
bmMl5s2,mMl5u1,mMl5u2,mMl5w1,mMl5w2,mMl5f1,mMl5f2,mMl5s1,
mMl5s2;. . .
    bmMl6u1,bmMl6u2,bmMl6w1,bmMl6w2,bmMl6f1,bmMl6f2,bmMl6s1,
bmMl6s2,mMl6u1,mMl6u2,mMl6w1,mMl6w2,mMl6f1,mMl6f2,mMl6s1,
mMl6s2;. . .
    bmMl7u1,bmMl7u2,bmMl7w1,bmMl7w2,bmMl7f1,bmMl7f2,bmMl7s1,
bmMl7s2,mMl7u1,mMl7u2,mMl7w1,mMl7w2,mMl7f1,mMl7f2,mMl7s1,
mMl7s2;. . .
    bmMl8u1,bmMl8u2,bmMl8w1,bmMl8w2,bmMl8f1,bmMl8f2,bmMl8s1,
bmMl8s2,mMl8u1,mMl8u2,mMl8w1,mMl8w2,mMl8f1,mMl8f2,mMl8s1,
mMl8s2];
```

```
Dl=[
zeros(1,8*N)                                              ; .  .  .

zeros(1,8*N)                                              ; .  .  .

zeros(1,8*N)                                              ; .  .  .

zeros(1,8*N)                                              ; .  .  .

zeros(1,8*N)                                              ; .  .  .

zeros(1,8*N)                                              ; .  .  .

zeros(1,8*N)                                              ; .  .  .

zeros(1,8*N)                                              ; .  .  .

zeros(1,8*N)                                              ; .  .  .

zeros(1,8*N)                                              ; .  .  .

zeros(1,8*N)                                              ; .  .  .

zeros(1,8*N)                                              ; .  .  .

zeros(1,8*N)                                              ; .  .  .

zeros(1,8*N)                                              ; .  .  .

zeros(1,8*N)                                              ; .  .  .

zeros(1,8*N)                                              ; .  .  .

zeros(1,8*N)                                              ; .  .  .

zeros(1,8*N)                                              ; .  .  .

zeros(1,8*N)                                              ; .  .  .
    bmDl1u1,bmDl1u2,bmDl1w1,bmDl1w2,bmDl1f1,bmDl1f2,bmDl1s1,
bmDl1s2,mDl1u1,mDl1u2,mDl1w1,mDl1w2,mDl1f1,mDl1f2,mDl1s1,
mDl1s2;. . .
    bmDl2u1,bmDl2u2,bmDl2w1,bmDl2w2,bmDl2f1,bmDl2f2,bmDl2s1,
bmDl2s2,mDl2u1,mDl2u2,mDl2w1,mDl2w2,mDl2f1,mDl2f2,mDl2s1,
mDl2s2;. . .
    bmDl3u1,bmDl3u2,bmDl3w1,bmDl3w2,bmDl3f1,bmDl3f2,bmDl3s1,
bmDl3s2,mDl3u1,mDl3u2,mDl3w1,mDl3w2,mDl3f1,mDl3f2,mDl3s1,
mDl3s2;. . .
    bmDl4u1,bmDl4u2,bmDl4w1,bmDl4w2,bmDl4f1,bmDl4f2,bmDl4s1,
bmDl4s2,mDl4u1,mDl4u2,mDl4w1,mDl4w2,mDl4f1,mDl4f2,mDl4s1,
```

```
mD14s2;. . .
    bmD15u1,bmD15u2,bmD15w1,bmD15w2,bmD15f1,bmD15f2,bmD15s1,
bmD15s2,mD15u1,mD15u2,mD15w1,mD15w2,mD15f1,mD15f2,mD15s1,
mD15s2;. . .
    bmD16u1,bmD16u2,bmD16w1,bmD16w2,bmD16f1,bmD16f2,bmD16s1,
bmD16s2,mD16u1,mD16u2,mD16w1,mD16w2,mD16f1,mD16f2,mD16s1,
mD16s2;. . .
    bmD17u1,bmD17u2,bmD17w1,bmD17w2,bmD17f1,bmD17f2,bmD17s1,
bmD17s2,mD17u1,mD17u2,mD17w1,mD17w2,mD17f1,mD17f2,mD17s1,
mD17s2;. . .
    bmD18u1,bmD18u2,bmD18w1,bmD18w2,bmD18f1,bmD18f2,bmD18s1,
bmD18s2,mD18u1,mD18u2,mD18w1,mD18w2,mD18f1,mD18f2,mD18s1,
mD18s2];
KLbb=Kl(1:20,1:20);
KLbd=Kl(1:20,20+1:end);
KLdb=Kl(20+1:end,1:20);
KLdd=Kl(20+1:end,20+1:end);
DLdb=Dl(20+1:end,1:20);
DLdd=Dl(20+1:end,20+1:end);
MLdb=Ml(20+1:end,1:20);
MLdd=Ml(20+1:end,20+1:end);

KL=KLdd-KLdb*inv(KLbb)*KLbd;
DL=DLdd-DLdb*inv(KLbb)*KLbd;
ML=MLdd-MLdb*inv(KLbb)*KLbd;

Np=(N-4);

Nmm=4*(N-2)+2*(N-4);

KLmm=KL(1:Nmm,1:Nmm);
KLme=KL(1:Nmm,Nmm+1:end);
KLem=KL(Nmm+1:end,1:Nmm);
KLee=KL(Nmm+1:end,Nmm+1:end);
DLmm=DL(1:Nmm,1:Nmm);
DLme=DL(1:Nmm,Nmm+1:end);
MLmm=ML(1:Nmm,1:Nmm);
MLme=ML(1:Nmm,Nmm+1:end);

KLnew=KLmm-KLme*inv(KLee)*KLem;
DLnew=DLmm-DLme*inv(KLee)*KLem;
MLnew=MLmm-MLme*inv(KLee)*KLem;

AAA1=[zeros(Nmm,Nmm),eye(Nmm);-inv(MLnew)*KLnew,-inv(MLnew)
*DLnew];

[qdL1,omg1]=eig(AAA1);

Nuu=N-2;
Nww=N-4;
```

```
u1ls=[0,qdL1(1:Nuu),0]';
u2ls=[0,qdL1(Nuu+1:2*Nuu),0]';
w1ls=[0,0,qdL1(2*Nuu+1:2*Nuu+Nww),0,0]';
w2ls=[0,0,qdL1(2*Nuu+Nww+1:2*Nuu+2*Nww),0,0]';
f1ls=[0,qdL1(2*Nuu+2*Nww+1:3*Nuu+2*Nww),0]';
f2ls=[0,qdL1(3*Nuu+2*Nww+1:4*Nuu+2*Nww),0]';

for i=1:size(omg1)
    omg2(i,1)=omg1(i,i);
end

omg2=(omg2*0.0012)*(sqrt(G/(eta1*rhot)));
omegim=abs(imag(omg2));
omegre=real(omg2);
omegims=sort(omegim);
omegres=sort(omegre);
freq1(ei,vi)=omegims(1);
freq2(ei,vi)=omegims(3);
freq3(ei,vi)=omegims(5);
freq4(ei,vi)=omegims(7);
damp11(ei,vi)=omegres(1);
damp12(ei,vi)=omegres(end);
damp21(ei,vi)=omegres(2);
damp22(ei,vi)=omegres(end-1);
damp31(ei,vi)=omegres(3);
damp32(ei,vi)=omegres(end-2);
damp41(ei,vi)=omegres(4);
damp42(ei,vi)=omegres(end-3);

u1lsm=u1ls;
w1lsm=w1ls;
f1lsm=f1ls;
u2lsm=u2ls;
w2lsm=w2ls;
f2lsm=f2ls;
%
error=1;
itter=0;
while freq1>1e2

    Knl1u1=zeros(N*N);
    Knl1u2=zeros(N*N);
    Knl1w1=a1*diag(C_1*w1lsm)*C_2;
    Knl1w2=zeros(N,N);
    Knl1f1=zeros(N,N);
    Knl1f2=zeros(N,N);
    Knl1s1=zeros(N,N);
    Knl1s2=zeros(N,N);
%    -----------------------------------------------
    Knl3u1=zeros(N,N);
    Knl3w1=a1*(diag(C_2*u1lsm)*C_1+diag(C_1*u1lsm)*C_2+1.5*
```

```
(diag(C_1*w11sm))^2*C_2);
    Knl3u2=zeros(N,N);
    Knl3w2=zeros(N,N);
    Knl3f1=zeros(N,N);
    Knl3f2=zeros(N,N);
    Knl3s1=3*gama*(diag(C_1*w11sm)*C_2+diag(C_2*w11sm)*C_1);
    Knl3s2=zeros(N,N);
%        ------------------------------------------------
    Knl5u1=zeros(N,N);
    Knl5w1=zeros(N,N);
    Knl5u2=zeros(N,N);
    Knl5w2=zeros(N,N);
    Knl5f1=zeros(N,N);
    Knl5f2=zeros(N,N);
    Knl5s1=zeros(N,N);
    Knl5s2=zeros(N,N);
%        --------------------------------------------------
    Knl7u1=zeros(N,N);
    Knl7w1=gama*diag(C_1*w11sm)*C_2;
    Knl7u2=zeros(N,N);
    Knl7w2=zeros(N,N);
    Knl7f1=zeros(N,N);
    Knl7f2=zeros(N,N);
    Knl7s1=zeros(N,N);
    Knl7s2=zeros(N,N);
%        --------------------------------------------------
    Knl2u1=zeros(N,N);
    Knl2u2=zeros(N,N);
    Knl2w1=zeros(N,N);
    Knl2w2=f2*diag(C_1*w21sm)*C_2;
    Knl2f1=zeros(N,N);
    Knl2f2=zeros(N,N);
    Knl2s1=zeros(N,N);
    Knl2s2=zeros(N,N);
%        ----------------------------------------------------
    Knl4u1=zeros(N,N);
    Knl4u2=zeros(N,N);
    Knl4w1=zeros(N,N);
    Knl4w2=f2*(diag(C_2*u21sm)*C_1+diag(C_1*u21sm)*C_2+1.5*
diag(C_1*w21sm)*C_2);
    Knl4f1=zeros(N,N);
    Knl4f2=zeros(N,N);
    Knl4s1=zeros(N,N);
    Knl4s2=3*gama*(diag(C_1*w21sm)*C_2+diag(C_2*w21sm)*C_1);
    %        ----------------------------------------------------
    Knl6u1=zeros(N,N);
    Knl6u2=zeros(N,N);
    Knl6w1=zeros(N,N);
    Knl6w2=zeros(N,N);
    Knl6f1=zeros(N,N);
    Knl6f2=zeros(N,N);
```

```
      Knl6s1=zeros(N,N);
      Knl6s2=zeros(N,N);
      %      ----------              --------------------------------
      Knl8u1=zeros(N,N);
      Knl8u2=zeros(N,N);
      Knl8w1=zeros(N,N);
      Knl8w2=gama*diag(C_1*w2lsm)*C_2;
      Knl8f1=zeros(N,N);
      Knl8f2=zeros(N,N);
      Knl8s1=zeros(N,N);
      Knl8s2=zeros(N,N);
%----------------------------------------
%----------------------------------------
bmKnl1u1=[Knl1u1(2:N-1,1),Knl1u1(2:N-1,N)];
bmKnl1u2=[Knl1u2(2:N-1,1),Knl1u2(2:N-1,N)];
bmKnl1w1=[Knl1w1(2:N-1,1:2),Knl1w1(2:N-1,N-1:N)];
bmKnl1w2=[Knl1w2(2:N-1,1:2),Knl1w2(2:N-1,N-1:N)];
bmKnl1f1=[Knl1f1(2:N-1,1),Knl1f1(2:N-1,N)];
bmKnl1f2=[Knl1f2(2:N-1,1),Knl1f2(2:N-1,N)];
bmKnl1s1=[Knl1s1(2:N-1,1),Knl1s1(2:N-1,N)];
bmKnl1s2=[Knl1s2(2:N-1,1),Knl1s2(2:N-1,N)];

bmKnl2u1=[Knl2u1(2:N-1,1),Knl2u1(2:N-1,N)];
bmKnl2u2=[Knl2u2(2:N-1,1),Knl2u2(2:N-1,N)];
bmKnl2w1=[Knl2w1(2:N-1,1:2),Knl2w1(2:N-1,N-1:N)];
bmKnl2w2=[Knl2w2(2:N-1,1:2),Knl2w2(2:N-1,N-1:N)];
bmKnl2f1=[Knl2f1(2:N-1,1),Knl2f1(2:N-1,N)];
bmKnl2f2=[Knl2f2(2:N-1,1),Knl2f2(2:N-1,N)];
bmKnl2s1=[Knl2s1(2:N-1,1),Knl2s1(2:N-1,N)];
bmKnl2s2=[Knl2s2(2:N-1,1),Knl2s2(2:N-1,N)];

bmKnl3u1=[Knl3u1(3:N-2,1),Knl3u1(3:N-2,N)];
bmKnl3u2=[Knl3u2(3:N-2,1),Knl3u2(3:N-2,N)];
bmKnl3w1=[Knl3w1(3:N-2,1:2),Knl3w1(3:N-2,N-1:N)];
bmKnl3w2=[Knl3w2(3:N-2,1:2),Knl3w2(3:N-2,N-1:N)];
bmKnl3f1=[Knl3f1(3:N-2,1),Knl3f1(3:N-2,N)];
bmKnl3f2=[Knl3f2(3:N-2,1),Knl3f2(3:N-2,N)];
bmKnl3s1=[Knl3s1(3:N-2,1),Knl3s1(3:N-2,N)];
bmKnl3s2=[Knl3s2(3:N-2,1),Knl3s2(3:N-2,N)];

bmKnl4u1=[Knl4u1(3:N-2,1),Knl4u1(3:N-2,N)];
bmKnl4u2=[Knl4u2(3:N-2,1),Knl4u2(3:N-2,N)];
bmKnl4w1=[Knl4w1(3:N-2,1:2),Knl4w1(3:N-2,N-1:N)];
bmKnl4w2=[Knl4w2(3:N-2,1:2),Knl4w2(3:N-2,N-1:N)];
bmKnl4f1=[Knl4f1(3:N-2,1),Knl4f1(3:N-2,N)];
bmKnl4f2=[Knl4f2(3:N-2,1),Knl4f2(3:N-2,N)];
bmKnl4s1=[Knl4s1(3:N-2,1),Knl4s1(3:N-2,N)];
bmKnl4s2=[Knl4s2(3:N-2,1),Knl4s2(3:N-2,N)];

bmKnl5u1=[Knl5u1(2:N-1,1),Knl5u1(2:N-1,N)];
bmKnl5u2=[Knl5u2(2:N-1,1),Knl5u2(2:N-1,N)];
```

```
bmKnl5w1=[Knl5w1(2:N-1,1:2),Knl5w1(2:N-1,N-1:N)];
bmKnl5w2=[Knl5w2(2:N-1,1:2),Knl5w2(2:N-1,N-1:N)];
bmKnl5f1=[Knl5f1(2:N-1,1),Knl5f1(2:N-1,N)];
bmKnl5f2=[Knl5f2(2:N-1,1),Knl5f2(2:N-1,N)];
bmKnl5s1=[Knl5s1(2:N-1,1),Knl5s1(2:N-1,N)];
bmKnl5s2=[Knl5s2(2:N-1,1),Knl5s2(2:N-1,N)];

bmKnl6u1=[Knl6u1(2:N-1,1),Knl6u1(2:N-1,N)];
bmKnl6u2=[Knl6u2(2:N-1,1),Knl6u2(2:N-1,N)];
bmKnl6w1=[Knl6w1(2:N-1,1:2),Knl6w1(2:N-1,N-1:N)];
bmKnl6w2=[Knl6w2(2:N-1,1:2),Knl6w2(2:N-1,N-1:N)];
bmKnl6f1=[Knl6f1(2:N-1,1),Knl6f1(2:N-1,N)];
bmKnl6f2=[Knl6f2(2:N-1,1),Knl6f2(2:N-1,N)];
bmKnl6s1=[Knl6s1(2:N-1,1),Knl6s1(2:N-1,N)];
bmKnl6s2=[Knl6s2(2:N-1,1),Knl6s2(2:N-1,N)];

bmKnl7u1=[Knl7u1(2:N-1,1),Knl7u1(2:N-1,N)];
bmKnl7u2=[Knl7u2(2:N-1,1),Knl7u2(2:N-1,N)];
bmKnl7w1=[Knl7w1(2:N-1,1:2),Knl7w1(2:N-1,N-1:N)];
bmKnl7w2=[Knl7w2(2:N-1,1:2),Knl7w2(2:N-1,N-1:N)];
bmKnl7f1=[Knl7f1(2:N-1,1),Knl7f1(2:N-1,N)];
bmKnl7f2=[Knl7f2(2:N-1,1),Knl7f2(2:N-1,N)];
bmKnl7s1=[Knl7s1(2:N-1,1),Knl7s1(2:N-1,N)];
bmKnl7s2=[Knl7s2(2:N-1,1),Knl7s2(2:N-1,N)];

bmKnl8u1=[Knl8u1(2:N-1,1),Knl8u1(2:N-1,N)];
bmKnl8u2=[Knl8u2(2:N-1,1),Knl8u2(2:N-1,N)];
bmKnl8w1=[Knl8w1(2:N-1,1:2),Knl8w1(2:N-1,N-1:N)];
bmKnl8w2=[Knl8w2(2:N-1,1:2),Knl8w2(2:N-1,N-1:N)];
bmKnl8f1=[Knl8f1(2:N-1,1),Knl8f1(2:N-1,N)];
bmKnl8f2=[Knl8f2(2:N-1,1),Knl8f2(2:N-1,N)];
bmKnl8s1=[Knl8s1(2:N-1,1),Knl8s1(2:N-1,N)];
bmKnl8s2=[Knl8s2(2:N-1,1),Knl8s2(2:N-1,N)];

%-------------------------------------------
%-------------------------------------------

mKnl1u1=Knl1u1(2:N-1,2:N-1);
mKnl1u2=Knl1u2(2:N-1,2:N-1);
mKnl1w1=Knl1w1(2:N-1,3:N-2);
mKnl1w2=Knl1w2(2:N-1,3:N-2);
mKnl1f1=Knl1f1(2:N-1,2:N-1);
mKnl1f2=Knl1f2(2:N-1,2:N-1);
mKnl1s1=Knl1s1(2:N-1,2:N-1);
mKnl1s2=Knl1s2(2:N-1,2:N-1);

mKnl2u1=Knl2u1(2:N-1,2:N-1);
mKnl2u2=Knl2u2(2:N-1,2:N-1);
mKnl2w1=Knl2w1(2:N-1,3:N-2);
mKnl2w2=Knl2w2(2:N-1,3:N-2);
mKnl2f1=Knl2f1(2:N-1,2:N-1);
```

```
mKnl2f2=Knl2f2(2:N-1,2:N-1);
mKnl2s1=Knl2s1(2:N-1,2:N-1);
mKnl2s2=Knl2s2(2:N-1,2:N-1);

mKnl3u1=Knl3u1(3:N-2,2:N-1);
mKnl3u2=Knl3u2(3:N-2,2:N-1);
mKnl3w1=Knl3w1(3:N-2,3:N-2);
mKnl3w2=Knl3w2(3:N-2,3:N-2);
mKnl3f1=Knl3f1(3:N-2,2:N-1);
mKnl3f2=Knl3f2(3:N-2,2:N-1);
mKnl3s1=Knl3s1(3:N-2,2:N-1);
mKnl3s2=Knl3s2(3:N-2,2:N-1);

mKnl4u1=Knl4u1(3:N-2,2:N-1);
mKnl4u2=Knl4u2(3:N-2,2:N-1);
mKnl4w1=Knl4w1(3:N-2,3:N-2);
mKnl4w2=Knl4w2(3:N-2,3:N-2);
mKnl4f1=Knl4f1(3:N-2,2:N-1);
mKnl4f2=Knl4f2(3:N-2,2:N-1);
mKnl4s1=Knl4s1(3:N-2,2:N-1);
mKnl4s2=Knl4s2(3:N-2,2:N-1);

mKnl5u1=Knl5u1(2:N-1,2:N-1);
mKnl5u2=Knl5u2(2:N-1,2:N-1);
mKnl5w1=Knl5w1(2:N-1,3:N-2);
mKnl5w2=Knl5w2(2:N-1,3:N-2);
mKnl5f1=Knl5f1(2:N-1,2:N-1);
mKnl5f2=Knl5f2(2:N-1,2:N-1);
mKnl5s1=Knl5s1(2:N-1,2:N-1);
mKnl5s2=Knl5s2(2:N-1,2:N-1);

mKnl6u1=Knl6u1(2:N-1,2:N-1);
mKnl6u2=Knl6u2(2:N-1,2:N-1);
mKnl6w1=Knl6w1(2:N-1,3:N-2);
mKnl6w2=Knl6w2(2:N-1,3:N-2);
mKnl6f1=Knl6f1(2:N-1,2:N-1);
mKnl6f2=Knl6f2(2:N-1,2:N-1);
mKnl6s1=Knl6s1(2:N-1,2:N-1);
mKnl6s2=Knl6s2(2:N-1,2:N-1);

mKnl7u1=Knl7u1(2:N-1,2:N-1);
mKnl7u2=Knl7u2(2:N-1,2:N-1);
mKnl7w1=Knl7w1(2:N-1,3:N-2);
mKnl7w2=Knl7w2(2:N-1,3:N-2);
mKnl7f1=Knl7f1(2:N-1,2:N-1);
mKnl7f2=Knl7f2(2:N-1,2:N-1);
mKnl7s1=Knl7s1(2:N-1,2:N-1);
mKnl7s2=Knl7s2(2:N-1,2:N-1);

mKnl8u1=Knl8u1(2:N-1,2:N-1);
mKnl8u2=Knl8u2(2:N-1,2:N-1);
```

```
mKnl8w1=Knl8w1(2:N-1,3:N-2);
mKnl8w2=Knl8w2(2:N-1,3:N-2);
mKnl8f1=Knl8f1(2:N-1,2:N-1);
mKnl8f2=Knl8f2(2:N-1,2:N-1);
mKnl8s1=Knl8s1(2:N-1,2:N-1);
mKnl8s2=Knl8s2(2:N-1,2:N-1);
%---------------------------------
%---------------------------------
bmKu11=bmKl1u1+bmKnl1u1;
bmKu12=bmKl1u2+bmKnl1u2;
bmKw11=bmKl1w1+bmKnl1w1;
bmKw12=bmKl1w2+bmKnl1w2;
bmKf11=bmKl1f1+bmKnl1f1;
bmKf12=bmKl1f2+bmKnl1f2;
bmKs11=bmKl1s1+bmKnl1s1;
bmKs12=bmKl1s2+bmKnl1s2;

bmKu21=bmKl2u1+bmKnl2u1;
bmKu22=bmKl2u2+bmKnl2u2;
bmKw21=bmKl2w1+bmKnl2w1;
bmKw22=bmKl2w2+bmKnl2w2;
bmKf21=bmKl2f1+bmKnl2f1;
bmKf22=bmKl2f2+bmKnl2f2;
bmKs21=bmKl2s1+bmKnl2s1;
bmKs22=bmKl2s2+bmKnl2s2;

bmKu31=bmKl3u1+bmKnl3u1;
bmKu32=bmKl3u2+bmKnl3u2;
bmKw31=bmKl3w1+bmKnl3w1;
bmKw32=bmKl3w2+bmKnl3w2;
bmKf31=bmKl3f1+bmKnl3f1;
bmKf32=bmKl3f2+bmKnl3f2;
bmKs31=bmKl3s1+bmKnl3s1;
bmKs32=bmKl3s2+bmKnl3s2;

bmKu41=bmKl4u1+bmKnl4u1;
bmKu42=bmKl4u2+bmKnl4u2;
bmKw41=bmKl4w1+bmKnl4w1;
bmKw42=bmKl4w2+bmKnl4w2;
bmKf41=bmKl4f1+bmKnl4f1;
bmKf42=bmKl4f2+bmKnl4f2;
bmKs41=bmKl4s1+bmKnl4s1;
bmKs42=bmKl4s2+bmKnl4s2;

bmKu51=bmKl5u1+bmKnl5u1;
bmKu52=bmKl5u2+bmKnl5u2;
bmKw51=bmKl5w1+bmKnl5w1;
bmKw52=bmKl5w2+bmKnl5w2;
bmKf51=bmKl5f1+bmKnl5f1;
bmKf52=bmKl5f2+bmKnl5f2;
bmKs51=bmKl5s1+bmKnl5s1;
bmKs52=bmKl5s2+bmKnl5s2;
```

```
bmKu61=bmKl6u1+bmKnl6u1;
bmKu62=bmKl6u2+bmKnl6u2;
bmKw61=bmKl6w1+hmKnl6w1;
bmKw62=bmKl6w2+bmKnl6w2;
bmKf61=bmKl6f1+bmKnl6f1;
bmKf62=bmKl6f2+bmKnl6f2;
bmKs61=bmKl6s1+bmKnl6s1;
bmKs62=bmKl6s2+bmKnl6s2;

bmKu71=bmKl7u1+bmKnl7u1;
bmKu72=bmKl7u2+bmKnl7u2;
bmKw71=bmKl7w1+bmKnl7w1;
bmKw72=bmKl7w2+bmKnl7w2;
bmKf71=bmKl7f1+bmKnl7f1;
bmKf72=bmKl7f2+bmKnl7f2;
bmKs71=bmKl7s1+bmKnl7s1;
bmKs72=bmKl7s2+bmKnl7s2;

bmKu81=bmKl8u1+bmKnl8u1;
bmKu82=bmKl8u2+bmKnl8u2;
bmKw81=bmKl8w1+bmKnl8w1;
bmKw82=bmKl8w2+bmKnl8w2;
bmKf81=bmKl8f1+bmKnl8f1;
bmKf82=bmKl8f2+bmKnl8f2;
bmKs81=bmKl8s1+bmKnl8s1;
bmKs82=bmKl8s2+bmKnl8s2;
%--------------------------------
mKu11=mKl1u1+mKnl1u1;
mKu12=mKl1u2+mKnl1u2;
mKw11=mKl1w1+mKnl1w1;
mKw12=mKl1w2+mKnl1w2;
mKf11=mKl1f1+mKnl1f1;
mKf12=mKl1f2+mKnl1f2;
mKs11=mKl1s1+mKnl1s1;
mKs12=mKl1s2+mKnl1s2;

mKu21=mKl2u1+mKnl2u1;
mKu22=mKl2u2+mKnl2u2;
mKw21=mKl2w1+mKnl2w1;
mKw22=mKl2w2+mKnl2w2;
mKf21=mKl2f1+mKnl2f1;
mKf22=mKl2f2+mKnl2f2;
mKs21=mKl2s1+mKnl2s1;
mKs22=mKl2s2+mKnl2s2;

mKu31=mKl3u1+mKnl3u1;
mKu32=mKl3u2+mKnl3u2;
mKw31=mKl3w1+mKnl3w1;
mKw32=mKl3w2+mKnl3w2;
mKf31=mKl3f1+mKnl3f1;
```

```
mKf32=mKl3f2+mKnl3f2;
mKs31=mKl3s1+mKnl3s1;
mKs32=mKl3s2+mKnl3s2;

mKu41=mKl4u1+mKnl4u1;
mKu42=mKl4u2+mKnl4u2;
mKw41=mKl4w1+mKnl4w1;
mKw42=mKl4w2+mKnl4w2;
mKf41=mKl4f1+mKnl4f1;
mKf42=mKl4f2+mKnl4f2;
mKs41=mKl4s1+mKnl4s1;
mKs42=mKl4s2+mKnl4s2;

mKu51=mKl5u1+mKnl5u1;
mKu52=mKl5u2+mKnl5u2;
mKw51=mKl5w1+mKnl5w1;
mKw52=mKl5w2+mKnl5w2;
mKf51=mKl5f1+mKnl5f1;
mKf52=mKl5f2+mKnl5f2;
mKs51=mKl5s1+mKnl5s1;
mKs52=mKl5s2+mKnl5s2;

mKu61=mKl6u1+mKnl6u1;
mKu62=mKl6u2+mKnl6u2;
mKw61=mKl6w1+mKnl6w1;
mKw62=mKl6w2+mKnl6w2;
mKf61=mKl6f1+mKnl6f1;
mKf62=mKl6f2+mKnl6f2;
mKs61=mKl6s1+mKnl6s1;
mKs62=mKl6s2+mKnl6s2;

mKu71=mKl7u1+mKnl7u1;
mKu72=mKl7u2+mKnl7u2;
mKw71=mKl7w1+mKnl7w1;
mKw72=mKl7w2+mKnl7w2;
mKf71=mKl7f1+mKnl7f1;
mKf72=mKl7f2+mKnl7f2;
mKs71=mKl7s1+mKnl7s1;
mKs72=mKl7s2+mKnl7s2;

mKu81=mKl8u1+mKnl8u1;
mKu82=mKl8u2+mKnl8u2;
mKw81=mKl8w1+mKnl8w1;
mKw82=mKl8w2+mKnl8w2;
mKf81=mKl8f1+mKnl8f1;
mKf82=mKl8f2+mKnl8f2;
mKs81=mKl8s1+mKnl8s1;
mKs82=mKl8s2+mKnl8s2;

%------------------------------------------------
```

```
Kll=[1 zeros(1,8*N-1);...
    0 1 zeros(1,8*N-2);...
    %---------------------------     ........
    zeros(1,2) 1 zeros(1,8*N-2-1);...
    zeros(1,2+1) 1 zeros(1,8*N-2*2);...
    %-----------------------------------
    zeros(1,2*2) 1 zeros(1,8*N-2*2-1);...
    zeros(1,2*2),Klw_2b,zeros(1,1*4),zeros(1,4*2),zeros
(1,2*(N-2)),Klw_2d,zeros(1,1*(N-4)),zeros(1,4*(N-2));...
    zeros(1,2*2),Klw_N1b,zeros(1,1*4),zeros(1,4*2),zeros
(1,2*(N-2)),Klw_N1d,zeros(1,1*(N-4)),zeros(1,4*(N-2));...
    zeros(1,2*2+3) 1 zeros(1,8*N-2*2-4);...
    %-----------------------------------
    zeros(1,2*2+1*4) 1 zeros(1,8*N-2*2-1*4-1);...
    zeros(1,2*2+1*4),Klw_2b,zeros(1,4*2),zeros(1,2*(N-2)),
zeros(1,1*(N-4)),Klw_2d,zeros(1,4*(N-2));...
    zeros(1,2*2+1*4),Klw_N1b,zeros(1,4*2),zeros(1,2*(N-2)),
zeros(1,1*(N-4)),Klw_N1d,zeros(1,4*(N-2));...
    zeros(1,2*2+1*4+3) 1 zeros(1,8*N-2*2-2*4);...
    %-----------------------------------
    zeros(1,2*2+2*4) 1 zeros(1,8*N-2*2-2*4-1);...
    zeros(1,2*2+2*4+1) 1 zeros(1,8*N-3*2-2*4);...
    %-----------------------------------
    zeros(1,3*2+2*4) 1 zeros(1,8*N-3*2-2*4-1);...
    zeros(1,3*2+2*4+1) 1 zeros(1,8*N-4*2-2*4);...
    %-----------------------------------
    zeros(1,4*2+2*4) 1 zeros(1,8*N-4*2-2*4-1);...
    zeros(1,4*2+2*4+1) 1 zeros(1,8*N-5*2-2*4);...
    %-----------------------------------
    zeros(1,5*2+2*4) 1 zeros(1,8*N-5*2-2*4-1);...
    zeros(1,5*2+2*4+1) 1 zeros(1,8*N-6*2-2*4);...
    %-----------------------------------
    bmKu11,bmKu12,bmKw11,bmKw12,bmKf11,bmKf12,mKu11,mKu12,
mKw11,mKw12,mKf11,mKf12;...
    bmKu21,bmKu22,bmKw21,bmKw22,bmKf21,bmKf22,mKu21,mKu22,
mKw21,mKw22,mKf21,mKf22;...
    bmKu31,bmKu32,bmKw31,bmKw32,bmKf31,bmKf32,mKu31,mKu32,
mKw31,mKw32,mKf31,mKf32;...
    bmKu41,bmKu42,bmKw41,bmKw42,bmKf41,bmKf42,mKu41,mKu42,
mKw41,mKw42,mKf41,mKf42;...
    bmKu51,bmKu52,bmKw51,bmKw52,bmKf51,bmKf52,mKu51,mKu52,
mKw51,mKw52,mKf51,mKf52;...
    bmKu61,bmKu62,bmKw61,bmKw62,bmKf61,bmKf62,mKu61,mKu62,
mKw61,mKw62,mKf61,mKf62...
    bmKu71,bmKu72,bmKw71,bmKw72,bmKf71,bmKf72,mKu71,mKu72,
mKw71,mKw72,mKf71,mKf72...
    bmKu81,bmKu82,bmKw81,bmKw82,bmKf81,bmKf82,mKu81,mKu82,
mKw81,mKw82,mKf81,mKf82];
Kbb=Kll(1:16,1:16);
Kbd=Kll(1:16,16+1:end);
```

```
Kdb=Kll(16+1:end,1:16);
Kdd=Kll(16+1:end,16+1:end);

KnL=Kdd-Kdb*inv(Kbb)*Kbd;

Np=(N-4);

Nmm=2*(N-2)+2*(N-4);

KnLmm=KnL(1:Nmm,1:Nmm);
KnLme=KnL(1:Nmm,Nmm+1:end);
KnLem=KnL(Nmm+1:end,1:Nmm);
KnLee=KnL(Nmm+1:end,Nmm+1:end);

Knew=KnLmm-KnLme*inv(KnLee)*KnLem;

    AAA2=[zeros(Nmm),eye(Nmm);-inv(MLnew)*Knew,-
inv(MLnew)*DLnew];
    [qdL1,omgnL1]=eig(AAA2);

    for i=1:size(omgnL1)
        omgnL2(i,1)=omgnL1(i,i);
    end
    error=(freq1-freq2)/freq1;

end
uf=uf+2.5/Q;
end
```

Index

Printed in the United States
by Baker & Taylor Publisher Services